CAVENDISH PROBLEMS IN CLASSICAL PHYSICS

compiled by

THE STAFF OF THE CAVENDISH LABORATORY, CAMBRIDGE

and edited by

A. B. PIPPARD

Sc.D., F.R.S.

Cavendish Professor of Physics
in the University of Cambridge and President of
Clare Hall, Cambridge

2nd edition revised by

W. O. SAXTON

Cavendish Laboratory, Cambridge

CAMBRIDGE

AT THE UNIVERSITY PRESS

1971

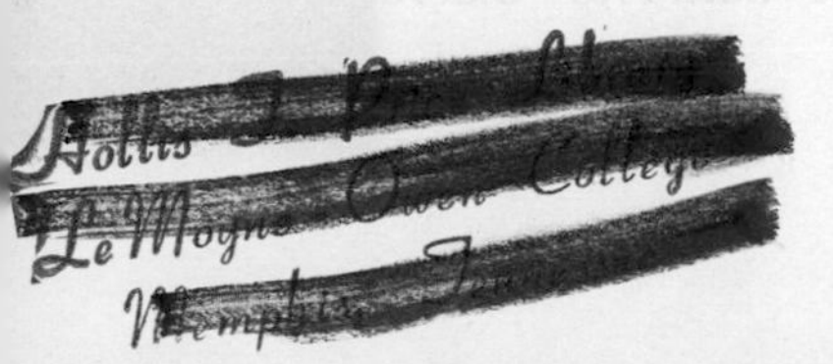

Published by the Syndics of the Cambridge University Press
Bentley House, 200 Euston Road, London NW1 2DB
American Branch: 32 East 57th Street, New York, N.Y.10022

ISBN: 0 521 08309 5

First edition 1962
Reprinted 1963
Second edition 1971

Printed in Great Britain by
Alden & Mowbray Ltd at the Alden Press, Oxford

CONTENTS

FROM PREFACE TO FIRST EDITION

Many of the problems in this book started as examination questions in Part I of the Natural Sciences Tripos, which is taken at the end of the second year at Cambridge. They have suffered some changes since then, and have been supplemented by specially invented problems, but the general level is the same. The university teacher, however, should not imagine that our purpose in publishing this collection is to provide him with a ready store of examination questions. We are much more concerned to help the serious student to understand physics, and it is his needs that we have tried to bear in mind throughout.

In recent years many attempts have been made to bring the teaching of physics up to date and to integrate with the still-valuable old ideas some of what is often quaintly called 'modern physics'. At a time when increasing numbers of students voluntarily embrace physics as a career, no effort is spared to make it seem even more attractive a prospect, and certainly we in Cambridge are trying to move with the times. Why, then, do we choose this moment to issue so old-fashioned-looking a compilation? is the Cavendish Laboratory still wedded to Heat, Light, Sound, Electricity and Magnetism? The answer is simple; it is precisely because we and so many others have moved away from the classical regimen that the need for this sort of book exists. No amount of exciting new presentations of the basic ideas is going to make it easy for the student to acquire professional competence. If he cannot make a good showing at these classical problems all his expertise in the realm of ideas is so much vanity; for they are examples of what will continually confront him in the practice of his art. And in fact they are by no means negligible as an intellectual pastime. There is much here to stimulate the imagination; in submitting oneself to an apparently humdrum exercise one may well discover a new way of looking at deeper problems. In the old curriculum there were many ideas taught which were of real value, yet not to such a degree that they force themselves into a modern course. A number of our problems are designed to encourage the student to think them out for himself, and so to see points of view he may never find in a text-book but which will enrich his appreciation of the more conventional subject-matter.

What we offer, then, is not the outline syllabus of a university course in physics; it is a set of exercises to develop the technique, like the five-finger exercises that are an indispensable basis of concert-hall mastery. They are intended for private study and for examples classes. In our attempt to display the variety of classical physics in a small compass we have found it impossible to provide a series of drill questions on each point of difficulty. But we have tried to expose every such difficulty so that the conscientious student may discover for himself where further reading is needed and which particular techniques he should practise further, by systematically working through those exercises with which modern text-books are so liberally endowed. If he has no difficulty with the majority of the questions, we recommend to his attention the starred questions which, though requiring no more knowledge, are rather harder in that they demand a more mature appreciation of principles.

It is my special duty and pleasure as editor to acknowledge with gratitude the help received from whoever they were who invented so many of the questions, from those present members of the Cavendish Laboratory who have given so much time to sifting and modifying the old and inventing the new, and from all the research students who have worked out the answers. I shall not pretend to believe that they are all correct, and shall be grateful to any reader who finds a mistake and lets me know.

A. B. PIPPARD

Cavendish Laborarory
Cambridge

PREFACE TO SECOND IMPRESSION

I have been notified of a number of errors by many friendly readers including, I am glad to say, some students; to all of these I extend my thanks. It is gratifying to find a second printing called for so soon, but it means that probably not all the errors have yet been detected, and I hope my critics will not relax their vigilance. As each new printing is prepared the necessary corrections will be made, and we hope to include additional problems as they are devised or brought to our attention. To avoid confusion each new printing will be given a differently coloured cover.

A. B. PIPPARD

PREFACE TO SECOND EDITION

This second edition takes advantage of the progress made in recent years towards international agreement on the names, symbols, and units to be used in connexion with physical quantities: the Système International d'Unités has been adopted, and the nomenclature and notation throughout have been revised in accordance with the new recommendations. The opportunity has been taken at the same time to correct a number of further errors, clarify some old problems, and insert a few new: it is hoped that the re-numbering thus necessitated will not cause serious inconvenience.

I am greatly indebted to the writers of a number of letters addressed to Professor Pippard pointing out errors and commenting on the problems: I hope more will write calling attention to the fresh errors this revision will inevitably have introduced. Some problems have been criticized on the grounds of the difficulty of obtaining an analytic solution: it is hoped, however, that these will encourage the student to develop graphical or numerical techniques.

W. O. SAXTON

SYMBOLS, UNITS AND DATA

All derived units of the Système International are obtained by combining the basic units without any numerical factors. Some of those with special names are listed below, as are some of the prefixes which may be used to vary thc size of the units. A definitive statement of the Système, together with recommended notational conventions, will be found in the report *Symbols, Signs, and Abbreviations* published in 1969 by the Royal Society. Attention should, however, be drawn here to two points:

(1) Indices apply to prefixes as well as to units,
i.e. $1\text{m} = 10^9\ \text{mm}^3$, not $10^3\ \text{mm}^3$;

(2) The mole remains the 'gramme molecule',
i.e. The Avogadro constant $= 6 \times 10^{23}\ \text{mol}^{-1}$, not $6 \times 10^{26}\ \text{mol}^{-1}$.

Quantity	*Unit*	*Abbreviation*
Basic units		
Distance	metre	m
Mass	kilogramme	kg
Time	second	s
Electric current	ampère	A
Thermodynamic temperature	kelvin	K
Amount of substance	mole	mol
Plane angle	radian	rad
Derived units		
Electric charge	coulomb	C
Capacitance	farad	F
Inductance	henry	H
Frequency	hertz	Hz
Energy	joule	J
Force	newton	N
Magnetic flux density	tesla	T
Electric potential	volt	V
Power	watt	W
Electric resistance	ohm	Ω

A number of familiar abbreviations for non-S.I. units are also used.

Prefixes

M-	mega-	$\times 10^6$	μ-	micro-	$\times 10^{-6}$
k-	kilo-	$\times 10^3$	n-	nano-	$\times 10^{-9}$
m-	milli-	$\times 10^{-3}$	p-	pico-	$\times 10^{-12}$

The familiar prefix c- (centi- $\times 10^{-2}$), now regarded as permissible but undesirable, will be found here in some units of volume.

Fundamental constants

Planck constant	$\hbar$	$= 1.055 \times 10^{-34}$ J s
Boltzmann constant	k	$= 1.381 \times 10^{-23}$ J K^{-1}
Gas constant	R	$= 8.31$ J K^{-1} mol^{-1}
Stefan–Boltzmann constant	σ	$= 5.67 \times 10^{-8}$ W m^{-2} K^{-4}
Gravitational constant	G	$= 6.67 \times 10^{-11}$ N m^2 kg^{-2}
Avogadro constant	N_A	$= 6.02 \times 10^{23}$ mol^{-1}
Permeability of a vacuum	μ_0	$= 4\pi \times 10^{-7}$ H m^{-1}
Speed of light in a vaccuum	c	$= 2.998 \times 10^8$ m s^{-1}
Charge of proton	e	$= 1.602 \times 10^{-19}$ C
Mass of electron	m_e	$= 9.11 \times 10^{-31}$ kg
Mass of proton	m_p	$= 1.673 \times 10^{-27}$ kg

$$[\varepsilon_0 = 1/\mu_0 c^2]$$

Other data

Gravitational acceleration (g)	$= 9.80$ m s^{-2} (nominal value)
Radius of the earth	$= 6370$ km
Density of mercury	$= 13.55$ Mg m^{-3}
Triple point of water	$= 273.16$ K (exact value)
1 atm	$= 101\,325$ N m^{-2} (exact value)
1 eV	$= 1.602 \times 10^{-19}$ J

PROBLEMS

1. A cyclist travelling on a straight road at 10 m.p.h. is subject to air resistance proportional to the square of his wind speed. By what factor must he increase his power output to maintain his speed if a 20 m.p.h. cross-wind develops?

2. A liner moves at 20 knots along a straight course that takes it within one nautical mile of a port. A tender capable of 12 knots leaves port at the last possible minute to intercept the liner. How far away is the liner when the tender leaves, and how far must the tender go to intercept it?

3. $ABCD$ is a regular tetrahedron with edges of unit length. Unit forces are applied along the edges AB and CD. Express their effect in terms of a force acting on the centre of the tetrahedron and a torque.

4. Two weightless rings slide on a smooth vertical circle and through the rings passes a smooth string which carries weights at the two ends, and at a point between the rings. If there is equilibrium when the rings are at points distant 30° from the highest point of the circle, find the relation between the three weights.

5. A uniform plank of thickness $2d$ and weight W is balanced horizontally across the top of a fixed circular cylinder of radius r, whose axis is horizontal and perpendicular to the length of the plank. Prove that the gain of potential energy when the plank is turned without slipping through an angle θ in a vertical plane is $W[r\theta\sin\theta-(r+d)(1-\cos\theta)]$ and deduce the condition of stability.

6. A rhombus $ABCD$ is formed of four equal uniform rods freely jointed together and suspended from the point A; it is kept in position by a light rod joining the midpoints of BC and CD; prove that if T be the thrust in this rod and W the weight of the rhombus, $T/W = BD/AC$.

7. A light ladder of length l is supported on a rough floor and leans against a rough vertical wall, the coefficient of friction having the same value μ at floor and wall. If the ladder is inclined at an angle α to the vertical, find how far a man can climb up the ladder before slipping begins.

8. A rope is wound round a fixed cylinder of radius r so as to make n complete turns. Show that if one end of the rope is held by a force F, a force $F\,e^{2\pi n\mu}$ must be applied to the other end to produce slipping, where μ is the coefficient of friction between rope and cylinder. Find also the work required to turn the cylinder through one complete turn under these conditions.

9. A uniform rope is suspended from two points 100 ft apart at the same level. If the rope hangs at 30° to the vertical at the points of attachment, how much lower is the middle of the rope, and how long is the rope?

10. Estimate roughly the magnitude of the following quantities:

(*a*) The total force exerted by a 40 m.p.h. gale on the side of a house 40 ft long and 20 ft high.

(*b*) The power required to keep in the air a helicopter of mass 500 kg whose blades are 3 m long. [Assume that all the air beneath the circle of the blades is moved uniformly downwards.]

(*c*) The work done in building the Great Pyramid whose base is 750 ft square and whose height is 500 ft. If all the power from a generating station delivering 60 MW could be used, how long would the job take?

(*d*) The height of the jet of a fountain which is supplied from a water main in which the absolute pressure is 3 atm.

11. Show that if kinetic energy is conserved in the collision of two bodies, it is conserved when the event is viewed in any frame of reference moving with constant velocity; and that if in one frame the kinetic energy is found to change, the same change will be found in all frames.

(*a*) An α-particle is projected with velocity 10^6 m s^{-1} directly at another α-particle which is initially at rest. What is the closest distance of approach and what is the ultimate behaviour of each particle?

(*b*) An α-particle collides elastically in a cloud chamber with the stationary nucleus of a helium atom and tracks are observed from both particles after the event. Show that the tracks are at right angles to one another.

(*c*) An α-particle collides elastically in a cloud chamber with a stationary proton. Show that the angle between the tracks is less than a right angle, and that the α-particle cannot be deflected through an angle greater than $14\frac{1}{2}°$.

(*d*) A nucleus of mass 20 a.m.u. is moving with a velocity of 3×10^6 m s^{-1} when it breaks into two fragments with an energy release of 10^{-12} J. The heavier fragment has mass 16 a.m.u. and is emitted at 90° to the original line of flight. What is the velocity of the lighter fragment?

[Mass of proton = 1 a.m.u. = 1.6×10^{-24} g.
Mass of α-particle = mass of helium nucelus = 4 a.m.u.
Charge on α-particle = 3.2×10^{-19} C.]

12. A jet aircraft at all speeds takes in 6000 lb of air per minute, burns 90 lb of fuel per minute and ejects the gases of combustion at a velocity of 1600 ft s^{-1} relative to the aircraft. Show how the thrust on the aircraft and the horse-power developed by the jet engine vary with the speed of the aircraft. Find the maximum thrust and horse-power and the maximum speed of the aircraft if air resistance can be ignored.

13. The body of a rocket has a mass m and in addition it contains initially a mass m_0 of fuel. Show that the velocity v of the rocket when the fuel is all burnt is given by

$$v = v_0 \log_e\left(1 + \frac{m_0}{m}\right),$$

where v_0 is the velocity of the ejected gases relative to the rocket. If $v_0 = 1000 \text{ m s}^{-1}$, $m_0 = 6$ tons and $m = 3$ tons, what is the maximum range of the rocket in the horizontal plane containing the position where the fuel is all burnt? Ignore the following effects: gravity during the burning period; air resistance; variation of gravity with height; curvature and motion of the earth.

14. A bucket of mass m is being drawn up a well by a rope which exerts a steady force F. Initially the bucket contains a mass m_0 of water, but this leaks out at a constant rate so that after time t the bucket is empty, before it reaches the top of the well. What is the velocity of the bucket when it is just empty?

15. A uniform chain is hung vertically with its lower end just touching a table top, and is then allowed to fall freely. Show that the force between the chain and the table increases quadratically with time, reaching a value three times the weight of the chain at the moment the upper end reaches the surface.

16. A large moving mass M makes a head-on elastic collision with a very small mass m, which is at rest. Find an expression for the fractional decrease of velocity of the large mass.

An artificial satellite of diameter 0.6 m and mass 80 kg travels in a circular orbit at 500 km above the earth's surface. What are its speed and period of revolution? The satellite's period is found to decrease by 1.85 s per day; calculate the corresponding drop in height and estimate roughly the density of the atmosphere at that height.

17. Calculate the ratio of the mean densities of the earth and the sun from the following approximate figures:

Angular diameter of sun seen from earth $= \frac{1}{2}°$.
Length of 1° of latitude on earth's surface = 100 km.
One year $= 10^{7.5}$ s.
$g = 10 \text{ m s}^{-2}$

18. According to the Bohr theory of the hydrogen atom, an electron of mass m and charge e under the influence of electrostatic attraction describes a circular orbit about a proton which may (except in section (e)) be assumed infinitely massive compared with the electron.

(*a*) Show that for any orbit the kinetic energy has a numerical magnitude half that of the potential energy, and that the total energy is the negative of the kinetic energy.

(*b*) If the angular momentum is permitted to take only values that are integral multiples of Planck's constant $\hbar$, show that the permitted values of the total energy are $-me^4/32\pi^2\varepsilon_0^2n^2\hbar^2$, where the quantum number n is any non-zero integer.

(*c*) Calculate the radius and total energy in electron volts of the smallest orbits ($n = 1$) in hydrogen and tin (nuclear charge = 50). [One electron volt is the kinetic energy of an electron after passage from rest through a potential difference of one volt.]

(*d*) When the electron changes from an orbit of quantum number n to one of $n-1$, ΔE, the energy released, is emitted as radiation of angular frequency $\Delta E/\hbar$. Show that as n increases this frequency approaches the angular frequency of the electron in the orbit.

(*e*) If the mass of the proton m_p is not taken as infinite, show that the above results should be corrected by replacing the electron mass m by a 'reduced mass' μ, where

$$\frac{1}{\mu} = \frac{1}{m} + \frac{1}{m_p}$$

(*f*) According to de Broglie, an electron moving with momentum p has associated with it a wave of wavelength $2\pi\hbar/p$. Show that Bohr's quantum condition (*b*) is equivalent to the requirement that the orbit perimeter be an integral number of wavelengths.

19*. An electron of mass m and charge e describes a closed orbit under the influence of a central force from the nucleus. Show that the magnetic moment due to the orbiting charge bears a constant ratio to angular momentum, irrespective of the form of the orbit, and that this ratio takes the value $e/2m$.

20* An electron moves with velocity v in a uniform magnetic field $\boldsymbol{B}$. Show that:

(*a*) the electron describes a helical path, rotating about an axis parallel to $\boldsymbol{B}$ with, angular velocity eB/m;

(*b*) if a uniform electric field $\boldsymbol{E}$ is now applied normal to $\boldsymbol{B}$, its effect is to add to the helical motion a constant velocity E/B normal to both $\boldsymbol{E}$ and $\boldsymbol{B}$.

21. What are the magnitude and direction of the deflexion caused by the Earth's rotation to:

(*a*) the bob of a plumb-line hung from the top to the bottom of the Eiffel Tower;

(*b*)* the point of impact of a body dropped from the top?

[The Eiffel Tower is 300 m high and is situated at latitude 49° N.]

22*. A point mass on the end of a light string rotates as a conical pendulum with angular velocity ω, the string being inclined at an angle θ to the vertical. Show that if the motion is slightly disturbed the resulting small oscillations have an angular frequency $\omega(1+3\cos^2\theta)^{\frac{1}{2}}$.

23. Show that the acceleration of a uniform sphere rolling down a plane inclined at an angle α is $\frac{5}{7}g \sin\alpha$.

24. A reel of thread whose radius is r, and moment of inertia about its axis mk^2, is allowed to unwind under gravity, the upper end of the thread being fixed. Find the acceleration of the reel and the tension in the thread.

25. A uniform heavy bar of length l is supported horizontally at rest. It is suddenly released and at the same instant it is struck a sharp blow vertically upwards, close to one end. Describe its subsequent motion. In a particular experiment it is observed to pass through its original position after a time t. Show that $t^2 = n\pi l/3g$, where n is an integer.

26. A door consists of a uniform rectangular panel of width b, height h and mass m, strengthened along all four edges by a rectangular frame, also of mass m, that is formed from material of small cross-section. The door is made to swing about a vertical axis along one edge. Where should a door-stop be placed if there is to be no reaction at the hinges when the door strikes it?

27. Two gear wheels, cut from the same sheet but with one twice the diameter of the other and having twice as many teeth, are mounted on parallel light axles far enough apart not to mesh. The larger wheel is spun with angular velocity ω and the wheels are then meshed. What is the subsequent angular velocity of the larger wheel, and what fraction of the original energy is lost?

28. The total mass of an aeroplane is 10^4kg and it has two landing wheels each of 2 m diameter and moment of inertia 100 kg m^2. The aeroplane touches down at 200 km per hour on a runway. Both wheels, which are quite free to turn but are not rotating before touchdown, touch down together and skid. Neglecting air resistance, calculate the speed of the aeroplane when skidding ceases.

29. A heavy disc free to rotate about its axis is mounted in a light framework. If the radius of the disc is 0.2 m and its angular velocity 300 rad s^{-1}, find its angular velocity of precession when supported with its axis horizontal at a point 0.15 m from the centre of mass of the disc.

30*. A coin of radius r spins on a smooth table with its plane at a small angle θ to the horizontal. Show that the head on the coin, viewed from above, appears to rotate with angular velocity $\sqrt{(g/r)}\theta^{\frac{3}{2}}$, where g is the acceleration due to gravity.

31*. A man in a free balloon is provided with a heavy wheel and axle and an electric motor which will drive it at great speed.

(*a*) The wheel with its axle vertical is initially at rest and is then set into rapid motion by the motor.

(*b*) After process (*a*) has been performed, the wheel is turned so that its axle is horizontal.

(*c*) After processes (*a*) and (*b*) have been performed the motor is switched off and the wheel comes gradually to rest.

What happens in each case? (Qualitative answers only.)

32*. An atomic nucleus may be considered to be a spinning rigid body with angular momentum $\boldsymbol{M}$ and magnetic moment $\gamma\boldsymbol{M}$. Show that when the nucleus is acted upon by a magnetic field $\boldsymbol{B}$ the axis of $\boldsymbol{M}$ precesses about $\boldsymbol{B}$, with an angular velocity $\gamma\boldsymbol{B}$, no matter what the angle between $\boldsymbol{B}$ and $\boldsymbol{M}$. If a small rotating magnetic field $\boldsymbol{b}$ is applied so that it is always normal to the plane containing $\boldsymbol{B}$ and $\boldsymbol{M}$, show that the angle between $\boldsymbol{B}$ and $\boldsymbol{M}$ changes steadily, and returns to its original value after a time $2\pi/(\gamma b)$. What is the behaviour if the rotational frequency of $\boldsymbol{b}$ is not exactly $\gamma\boldsymbol{B}$?

33*. A uniform plank stands on a smooth floor vertically against a smooth wall; its lower end is pushed gently so as to slide freely away from the wall. Show that when it makes an angle $\sin^{-1}\frac{2}{3}$ with the horizontal its upper end leaves the wall.

34. A small body rests on a horizontal diaphragm of a loudspeaker which is being supplied with an alternating current of constant amplitude but the frequency of which can be varied. If the overall movement of the diaphragm is 10 μm at all frequencies, find the greatest frequency for which the small body stays in contact with the diaphragm.

35. A pendulum consists of a uniform circular disc of radius r free to turn about a horizontal axis perpendicular to its plane. Find the position of the axis for which the periodic time is a minimum and show how the value of the acceleration due to gravity would be obtained from the minimum periodic time.

36. A disc is suspended with its plane horizontal by three equidistant parallel strings, each of length l, attached to points on the circumference. Show that when it is displaced through a small angle about a vertical axis through its centre and released, its time of vibration is equal to that of a simple pendulum of length $l/2$.

37. A uniform horizontal disc of mass m and radius r is suspended at its centre from a vertical torsion wire and performs simple harmonic angular oscillations with period T_0, amplitude A_0, and energy E_0. A wire ring of mass m and radius $\frac{1}{2}r$ is dropped on to the disc and at once sticks to the disc.

Discuss what happens to (*a*) the period, (*b*) the amplitude, and (*c*) the energy of the oscillation in the two cases where the ring is dropped on (i) at the end of the swing when the disc is at rest, and (ii) at the centre of the swing when the disc is moving with its maximum velocity. Assume that the disc and ring are concentric.

38. A body whose moment of inertia is I hangs from a wire for which a torque T is required to turn the ends relative to one another by 1 rad. The upper end of the wire is suddenly twisted through an angle θ. When the body has reached the limit of its swing, the wire is suddenly twisted back to its original position. Derive the angular velocity of the body as it passes through its original position.

39. A body of moment of inertia I is suspended from a torsion fibre for which the restoring torque per unit angular displacement is T; when the angular velocity of the body is Ω it experiences a retarding torque $k\Omega$. If the top end of the fibre is made to oscillate with angular displacement $\phi = \phi_0 \sin\omega t$, where $\omega^2 = T/I$, show that the maximum twist in the wire is $\phi_0(1+TI/k^2)^{\frac{1}{2}}$.

40. A mass suspended from a spring and subject to a viscous damping force is acted upon by a periodic force of constant amplitude whose frequency ω is variable. Show that if the damping is small the curve representing the square of the amplitude of oscillation, y, as a function of ω is to a good approximation symmetric about the resonant frequency ω_0, according to the law

$$y/y_0 = a^2/(a^2+\delta^2),$$

in which y_0 is the square of the amplitude at resonance, a is a measure of the damping, and $\delta = \omega - \omega_0$.

The Quality Factor, Q, may be defined in the following ways:

(*a*) The width of the resonance, i.e. the difference $\Delta\omega$ between the two frequencies which make $y = \frac{1}{2}y_0$, is ω_0/Q.

(*b*) If the oscillation is allowed to decay freely, the fraction of the energy of oscillation dissipated per cycle is $2\pi/Q$.

(*c*) The amplitude of response at resonance is Q times the movement that would be imparted to the mass by the same force applied steadily.

Show that if Q is large all these definitions are equivalent.

41. If a periodic force, represented by $Fe^{i\omega t}$, acts upon a body whose resulting velocity is $ve^{i\omega t}$, F and v being in general complex numbers, show that the mean rate of energy transfer to the body is one-half the real part of Fv^*, v^* being the complex conjugate of v.

Use the model of a resonant system described in question 40 to show that at each frequency of the applied force the phase and amplitude of the oscillation adjust themselves so that the energy supplied by the force equals the energy dissipated by viscosity.

42. If the system in question 40 is at rest when the periodic force is first applied, show that about $1.5Q$ cycles must elapse before the amplitude of oscillation has settled down within 1% of its steady value.

43. The two electrons in the helium atom (atomic weight 4) may be assumed to be capable of vibration about their mean positions with a frequency which can be roughly determined from the observation that helium gas heavily absorbs ultra-violet radiation of wavelength 59 nm. Estimate the static dielectric constant of liquid helium whose density is 140 kg m^{-3}.

44. A mass hangs from a spring which exerts a force μx when stretched by an amount x. The top of the spring is attached to a beam which is deflected αW by a load W. Show that the mass oscillates on the spring as if the force constant were not μ but $\mu/(1+\alpha\mu)$.

45*. A simple pendulum consisting of a mass suspended from a string of length l suffers viscous damping such as to reduce its amplitude by a small fraction α in each cycle. The string is attached to a support which is caused to oscillate vertically with amplitude a at twice the natural frequency of the pendulum. Show (e.g. by considering the energy balance) that if a exceeds a critical value $\alpha l/2\pi$ the pendulum will be stimulated to swing with a large amplitude.

46*. Two identical springs, carrying equal masses, are hung from the beam of question 44. Show that if the masses are set oscillating with the same amplitude either in phase or in antiphase their motions are simple harmonic, but with frequencies in the ratio $(1+2\alpha\mu)^{\frac{1}{2}}$. Show also that if one mass is set oscillating while the other is stationary, the energy of oscillation is periodically transferred from one mass to the other, the period of the transfer process being approximately $1/(\alpha\mu)$ cycles of each oscillator if α is small.

47*. A large number of identical masses m, arranged in a line at equal intervals a, are joined together by identical springs between neighbours, the springs being such that unit extension requires a force μ. The mass at one end is oscillated along the direction of the line with angular frequency ω. Show that a compressional wave is propagated along the line with wave-number k ($= 2\pi/\lambda$) given by the expression

$$\omega/\omega_0 = \sin(\tfrac{1}{2}ka), \quad \text{where} \quad \omega_0 = 2(\mu/m)^{\frac{1}{2}}.$$

What happens if ω is made greater than ω_0?

48. Show that the kinetic and potential energies are equal in a sinusoidal wave travelling on a stretched string.

49. What is the ratio of the velocities of transverse and longitudinal waves on a wire which is stretched elastically by 0.1%?

50. A uniform inextensible string of length l and total weight W is suspended vertically and tapped at the top end so that a transverse impulse runs down it. At the same moment a body is released from rest and falls freely from the top of the string. How far from the bottom does the body pass the impulse? What weight should be hung from the string so that both reach the bottom together?

51*. A very long inextensible string is stretched horizontally under uniform tension by being passed over a pulley to which is attached a weight. The other end is fixed to a support which may be harmonically oscillated in a vertical line. Show that:

(*a*) the work needed to make the support oscillate equals the energy of the wave which is thus generated;

(*b*) the weight rises steadily and the momentum of the horizontal part of the string increases;

(*c*) hence the momentum in a given length of the wave is one-half the energy divided by the wave velocity;

(*d*) the mean horizontal component of the force at the oscillating end is less than that at the pulley by the right amount to supply the momentum of the wave.

52*. Calculate the fraction of power reflected at a sharp boundary in each of the following cases, when a wave passes from the first medium into the second (for the sound waves assume normal incidence of a plane wave on a plane boundary). State also whether the standing wave set up in the first medium has a maximum or minimum amplitude at the boundary.

(*a*) A string of mass 10 g m^{-1}, is joined to one of mass 40 g m^{-1} and the composite string is put under constant tension and oscillated transversely.

(*b*) Sound waves pass from water ($v = 1.54$ km s^{-1}) into ice ($v = 2.1$ km s^{-1}, $\rho = 0.92$ Mg m^{-3}).

(*c*) Longitudinal waves in a rod of aluminium ($E = 7.05 \times 10^{10}$N m^{-2}, $\rho = 2.7$ Mg m^{-3}) pass into lead (E $= 1.62 \times 10^{10}$ N m^{-2}, $\rho =$ 11.3 Mg m^{-3}).

(*d*) Sound waves pass from nitrogen (molecular weight $= 28$, $\gamma = \frac{7}{5}$) into helium (molecular weight $= 4$, $\gamma = \frac{5}{3}$).

53. A string of mass ρ per unit length is held under tension F. One end is so arranged that it is free to move transversely, but being attached to a light vane immersed in a viscous liquid it experiences a force αw proportional to the transverse velocity w. Show that if α is very small or very large a wave incident on the end is totally reflected, producing an antinode at the end in the first case and a node in the second; determine the value of α for which there is no reflexion.

54. A string of mass 200 g m^{-1} has a point mass of 5 g attached to it and is put under uniform tension. What fraction of the energy in a wave of wavelength 0.1 m is reflected by the point mass?

* At what distance from the first mass should another identical mass be attached so as to eliminate the reflexion?

55. Show that in dispersive wave-motion the wave velocity v and group velocity u are related by the following equivalent formulae:

(*a*) $v = \omega/k$, $u = \mathrm{d}\omega/\mathrm{d}k$,
(*b*) $u = v - \lambda \mathrm{d}v/\mathrm{d}\lambda$,
(*c*) $u = v\left(1 + \dfrac{\lambda}{n}\dfrac{\mathrm{d}n}{\mathrm{d}\lambda}\right)$,

where ω = angular frequency, k = wave-number = $2\pi/\lambda$, λ = wave-length measured in the medium, n = refractive index.

56. The velocity in air of light of wavelength 0.5 μm is determined by timing a pulse of light over a known distance, and found to be 299 712.6 km s^{-1}. The refractive index of air varies with vacuum wavelength λ according to the formula.

$$n-1 = A + B/\lambda^2, \quad \text{with} \quad A = 2.726\times10^{-4},\ B = 1.54\ \mathrm{nm}^2.$$

What is the free space velocity of light?

57. The wave velocity of surface waves on a deep liquid is given by the expression

$$v^2 = g/k + \gamma k/\rho,$$

where g = gravitational acceleration, k = wave-number $2\pi/\lambda$, γ = surface tension, ρ = density. Determine the ratio of group velocity to wave velocity when the dominant mechanism is (*a*) gravity, (*b*) surface tension, and estimate the minimum value of the group velocity.

58. In wave-mechanics a wave is associated with a particle, and in order for this association to be reasonable the following conditions must be satisfied:

(i) when the particle crosses a plane boundary separating two regions in which its potential energy is different, the wavelength must change so that both particle and wave obey the same law of refraction;

(ii) the group velocity of the wave must be the same as the velocity of the particle.

Show that if the frequency ν and wavelength λ of the wave are related to the total energy E and momentum p of the particle by de Broglie's relations,

$$\nu = E/h, \quad \lambda = h/p,$$

where h is a constant (Planck's constant), then the above conditions are satisfied.

Show also that if the particle is reflected at normal incidence from a moving mirror, the change in E and the change in frequency of the wave due to the Doppler effect are in accord with de Broglie's relations.

59. The D-lines from atomic sodium (atomic weight 23) in a vapour lamp are observed in a high resolution spectrometer each to have a breadth of about 4 pm. Taking the wavelength of the lines to be 0.6 μm estimate the temperature of the source, on the assumption that the broadening is due entirely to the Doppler effect.

60. A rocket equipped with a radio transmitter whose frequency is accurately controlled at 10 MHz is sent into the ionized regions of the upper atmosphere where the refractive index is known to be less than unity. When the velocity of recession in the line of sight is 600 m s^{-1} the received signal, mixed with another 10 MHz oscillation, gives a beat frequency of 10 Hz. What is the refractive index of the medium in which the rocket is moving?

61*. An artificial satellite emitting a constant frequency radio signal passes by an observing station which records the variation of received frequency at 20 s intervals as follows: 40.002 15, 40.002 08, 40.001 96, 40.001 75, 40.001 41, 40.001 06, 40.000 77, 40.000 59, 40.000 49 40.000 43 MHz.

Assuming a straight trajectory for the satellite calculate its speed and closest distance of approach.

62. A wave is generated inside a perfectly reflecting sphere which is subsequently caused to expand slowly. Show that the Doppler effect causes the wavelength to increase in proportion to the radius of the sphere. Hence show that a standing wave pattern established within the sphere is not changed in form by uniform slow expansion.

63. An observer standing a distance D from a fence whose vertical palings are equally spaced by an amount a claps his hands once. Describe the sound he hears by reflexion from the fence.

64. A cubic box of volume V has rigid walls and contains a gas in which the velocity of sound is v, independent of frequency. What is the lowest resonant frequency of the box? Show that the possible resonances have frequencies which are constant multiples of $(l^2+m^2+n^2)^{\frac{1}{2}}$, where l, m and n are integers. Hence show that the number of different resonances with frequency less than ν is approximately $\frac{4}{3}\pi V\nu^3/v^3$.

65. Use Huygens's construction to demonstrate the following results:

(a) If the refractive index of a material varies according to the law $n = n_0+\alpha x$, a ray which is nearly parallel to the y-axis is bent into an arc of radius n/α.

(*b*) If a ray of light in free space falls at an angle θ to the normal of a plane mirror which is receding with velocity v ($\ll c$), the ray is reflected at an angle ϕ such that

$$\phi - \theta \approx \frac{2v}{c} \sin\, \theta.$$

66. A disc of radius 1 m and thickness 0.25 m is constructed for use as a microwave aerial out of an artificial dielectric whose refractive index varies from 1.5 at the centre to 1.1 at the edge, but is constant throughout the thickness at any point. How should the refractive index vary with radius for the disc to act as a convex lens, and what will be its focal length? It may be assumed that the focal length is much greater than 1 m.

67. A capillary tube has internal diameter d and is made of glass of refractive index n. What is the apparent internal diameter when the tube is viewed from the side?

68. The lowest component of an oil-immersion microscope objective is a hemisphere of radius 4.5 mm with its plane surface immersed in oil. If the refractive indices of lens, oil and cover-slip are all 1.515, how far below the plane surface should the object be placed for aplanatic refraction to occur?

69*. Two identical thin convex lenses are fixed a certain distance apart, and measurements are made of the positions of object and real image, as given in the table below. The distances are measured on opposite sides from the same point. Determine the focal length and separation of the lenses.

Object distance	18.1	19.4	21.2	22.7	24.7	25.9	27.1	29.8	33.8
Image distance	46.2	42.3	38.4	35.9	33.3	32.1	31.0	29.1	27.0

70. At the focus of a parabolic mirror of radius 0.1 m and focal length 1 m is placed a thin matt disc which is just large enough to receive the image of the sun formed by the mirror. If the sun radiates as a black body at 6000 K, what is the highest temperature to which it can raise the disc?

71. What is the separation of the interference fringes produced by a biprism with angles of $1° \ 30'$ and index of refraction 1.52, if the light is monochromatic with wavelength 0.66 μm and the distances from source slit to biprism and biprism to screen are 0.2 m and 0.8 m respectively? How many fringes could you hope to see on the screen?

72. In an experiment to demonstrate Young's fringes, coherent light from a source slit falls on two narrow slits which are 1 mm apart and 0.1 m from the source; the fringes are observed on a screen 1 m away. The source is white light filtered so that only the wavelength band 0.48 to 0.52 μm is used.

(*a*) What is the separation of the fringes on the screen?

(*b*) Roughly how many fringes will be clearly visible?

(*c*)* If in front of one slit is placed a glass slide 0.20 mm thick, having refractive index 1.664 at 0.48 μm and 1.656 at 0.52 μm, how far will the centre of the pattern shift?

(*d*) How wide can the source slit be made without serious impairment of the fringe visibility?

73. Newton's rings are formed by reflexion in the air film between a plane surface and a spherical surface of radius 0.5 m. If the diameter of the third bright ring is 1.81 mm and of the twenty-third 5.01 mm, what is the wavelength of the light used?

74. A pair of accurately parallel half-silvered plates is illuminated by an extended monochromatic source and viewed through a telescope. A system of circular fringes is observed and the angular radii of the first ten are measured to give the following values: 2° 28′, 6° 22′, 8° 42′, 10° 24′, 11° 57′, 13° 18′, 14° 33′, 15° 42′, 16° 45′, 17° 45′. Calculate as accurately as possible the ratio of the plate separation to the wavelength and estimate the probable error of your result.

75. Monochromatic light of wavelength λ emerging from a pinhole passes through a parallel-sided glass plate, of thickness d and refractive index n, placed a distance R_1 from the pinhole; it then falls on a screen parallel to the plate and R_2 from it. Show that circular fringes appear on the screen. If a bright fringe appears at the centre determine the radii of subsequent bright fringes, on the assumption that their angular radii are small.

* If the parallel plate is replaced by one of the same mean thickness but very slightly tapered, show that in first approximation the fringe system is unchanged in form but shifted a distance $R_1(R_1+R_2)n^2\theta/d$, θ being the angle of taper of the plate.

76*. A perfect lens forms an image of a transparent object illuminated from behind by light of wavelength λ from a distant source. The optical thickness of the object varies across its surface, undergoing small sinusoidal oscillations about its mean value. Show that corresponding variations in intensity will be observed in general on a screen placed a small distance d from the true image plane, but that the variation will vanish if its spatial frequency is $(n/\lambda d)^{\frac{1}{2}}$, where n is any integer.

77. Two pinholes, of diameter 0.05 mm and 2.0 mm apart, are illuminated by parallel monochromatic light of wavelength 0.6 μm. A convex lens of diameter 10 mm and focal length 0.5 m is placed 0.6 m from the holes. What will be seen (*a*) 0.5 m (*b*) 3 m behind the lens?

78. A narrow slit is placed 1 m from a convex lens of focal length 0.5 m and a sharp image of it is formed on a screen with light of wavelength 0.5 μm. There is now interposed in the beam a second slit parallel to the first and 0.5 mm wide. What is the width between the first zero on each side of the maximum of the pattern on the screen if the second slit is placed (*a*) near the lens, (*b*) half way between the first slit and the lens, (*c*) half way between the lens and the screen?

79. Light of wavelength 0.5 μm from a point source is made into a parallel beam by means of a lens and then with a second lens of 1 m focal length is focused on to a screen. Various diffracting objects are placed between the lenses. State where maxima of intensity are to be found on the screen for the following objects:

(*a*) an opaque sheet with a long slit 0.5 mm wide cut in it;

(*b*) a wire 0.5 mm in diameter.

State where zero intensity occurs for the following holes pierced in an opaque sheet:

(*c*) a rectangular hole 5×0.5 mm;

(*d*) three small circular holes 0.5 mm apart in a line;

(*e*)* three small circular holes at the apices of an equilateral triangle of side 1 mm.

80. Monochromatic light falls, not at normal incidence, on a grating of 1200 lines/mm and two successive diffracted beams are observed at 73° and 14° to the normal on the same side. What is the wavelength of the light, and in what other directions should diffracted beams be seen?

81. A grating 50 mm long is ruled with 200 lines/mm, the opaque portions being twice as wide as the transparent portions, and is illuminated normally with plane monochromatic light of wavelength 0.5 μm. At what angles do the first five diffracted beams appear, what are their intensities relative to the central transmitted beam, and what, approximately, is their angular width?

82. Monochromatic light falls normally on a regular grating which is so ruled that the central beam and the diffracted beams all have the same intensity, taken as unity. The first diffracted beam is at 4° to the normal. If every fifth ruling of the grating is now obscured, calculate the positions and intensities of all beams up to 8° from the normal.

83. In a Rayleigh refractometer the gas tube is 150.23 mm long. It is observed that when sodium light (λ = 0.5893 μm) is used a change of air pressure from 758.4 mmHg to 49.1 mmHg causes 70.00 ± 0.05 fringes to pass the cross-wire. Determine the refractive index of air at a pressure of one atmosphere and at the temperature of the experiment; give the probable error of the result.

84. Estimate the smallest angular diameter of a star which could be recognized as other than a point when observed through a telescope whose objective is 100 inches in diameter, if it is used (*a*) conventionally, (*b*) as part of a Michelson stellar interferometer with mirrors up to 30 ft apart.

85. You are provided with a telescope object lens 0.1 m in diameter and of focal length 3 m. Given that the eye can just resolve objects separated by one minute of arc, what should the focal length of the eye-piece be to make full use of the objective? At what distance could you read this book with the aid of such a telescope?

86. How many times could you (in principle) write the Lord's Prayer on a flat 1 mm pinhead and still be able to read the letters with the aid of a good microscope?

87. The sodium *D*-lines have wavelengths 0.5890 μm and 0.5896 μm. Determine the length of side of a 60° glass prism which could just resolve these lines at minimum deviation, if the refractive index varies with wavelength as follows:

λ	0.5461	0.5893	0.6438 μm
n	1.6546	1.6499	1.6434

How many lines in all must a diffraction grating have, and how many per mm, to give the same resolution and angle of deviation in the first-order spectrum at normal incidence?

88*. A Fabry–Perot interferometer is made by coating each face of a plane parallel slab of fused silica (refractive index 1.46) so that it reflects 95% and transmits 4% of the light intensity incident on it. If the slab is 10 mm thick, what is the closest doublet of mean wavelength 0.6 μm that can be resolved? At the interference maxima what is the transmitted intensity as a fraction of the incident intensity? At what angle to the normal does the tenth fringe from the centre emerge?

89. A point source of monochromatic light of wavelength 0.5 μm is placed 0.4 m away from a circular aperture of radius 1 mm. At what distances from the aperture will it produce on a screen a pattern having a dark centre?

90. The Newton's rings of question 73 are photographed with unit magnification. Show that a transparency of this photograph can act as a zone plate and determine its principal focal length at the same wave-length. A lens of crown glass is placed in contact with the zone plate and chosen so that the combination is free from chromatic aberration so far as the principal focus is concerned. If the refractive index of the glass is

1.500 at a wavelength of 0.8 μm and 1.535 at 0.4 μm what is the focal length of the combination?

91*. Light of wavelength 0.53 μm from a pinhole passes through a slit 1m away, and thence falls on a screen 0.5 m behind the slit. How wide should the slit be for the diffraction pattern to have two well-marked maxima? Determine approximately the separation of the maxima.

92*. Microwave radiation of wavelength 30 mm is radiated in a wide angle from a horn and falls on a large perfectly reflecting plane surface 0.6 m away, so that a fraction of the power is reflected back into the horn. Show that by cutting out a circular disc from the reflecting surface immediately opposite the horn, and moving this disc a certain distance towards the horn, the power reflected into the horn can be reduced to zero. Determine the diameter of the smallest circle that will achieve this result and how it must be moved.

This device is sometimes used in microwave aerials in which the horn is placed at the focus of a reflecting paraboloid of revolution. Show that if the focal length of the paraboloid is 0.6 m the power reflected back is four times as much as from a plane 0.6 m from the horn, and determine what diameter must be the circular cap cut from the pole and how far it must be moved to annul the reflected power in this case.

93. A horizontal parallel beam of light, known to be elliptically polarized, is passed through first a quarter-wave plate and then a Nicol prism. It is found that for a certain orientation of the quarter-wave plate the emergent light is plane polarized at 23° to the vertical and that when the quarter-wave plate is turned through a right angle it is plane polarized at 81° to the vertical. Determine the ratio of the axes of the ellipse and the inclination of the major axis.

94. A parallel beam of white light passes through crossed Nicols between which is a plate of quartz 44 mm thick, and then into a spectrometer where it is observed to form a banded spectrum with dark bands at the wavelengths 0.434, 0.456, 0.482, 0.513, 0.551, 0.599 μm. Show that this is consistent with a rotation of the plane of polarization which for a given thickness varies with wavelength as $a+b/\lambda^2$, a and b being constants. If at 0.5893 μm the rotation of quartz is 21.24 ° mm^{-1}, find the thickness which will produce a rotation of one revolution in light of 0.5 μm.

95. A beam of plane polarized light of wavelength 0.5 μm is passed through a sugar solution which casues the plane of polarization to rotate through 9° mm^{-1}. What is the difference in the refractive indices for the two senses of circular polarized light?

After traversing 1m distance in the solution the light is found to have become, through a difference in absorption of the two circular components, elliptically polarized with an axial ratio of 3. What will the axial ratio be after 2 m?

96. A narrow beam of light enters a uniaxial crystal at grazing incidence in a direction at right angles to the optic axis, which is parallel to the face of the crystal. If the crystal is 50 mm thick and the refractive indices for ordinary and extraordinary rays are 1.525 and 1.479, what is the separation of the beams on the opposite face?

97*. A beam of plane polarized monochromatic light is passed through a quarter-wave plate which converts it to circularly polarized light, and then through a half-wave plate and a second quarter-wave plate. The half-wave plate is now rotated about an axis normal to its plane at a speed of 6000 r.p.m. Show that what emerges from the second quarter-wave plate is plane polarized light having a different frequency from the light entering the system, and determine by how much the frequency is altered.

98. The differential equation for the amplitude ψ of the electron wave in the quantum theory of the hydrogen atom takes the form

$$\frac{\hbar^2}{2m_e}\nabla^2\psi+\left(E+\frac{e^2}{4\pi\varepsilon_0 r}\right)\psi = 0,$$

in which ∇^2 is the differential operator $\partial^2/\partial x^2+\partial^2/\partial y^2+\partial^2/\partial z^2$,

$$r = (x^2+y^2+z^2)^{\frac{1}{2}},$$

E is the energy of the electron, and $\hbar$, m_e, e and ε_0 are constants whose values will be found on p. x.

Show that if the system of units is changed so that $\hbar$, m, e and ε_0 are taken as units of their respective quantities, the equation takes the form

$$\nabla^2\psi+2\left(E+\frac{1}{4\pi r}\right)\psi = 0.$$

How long is a metre and what is the velocity of light in this system? How many electron volts are there in the unit of energy?

99. In a proposed rational system of mechanical units the following fundamental constants are defined to have the value unity (and are therefore dimensionless): the ratio of gravitational to inertial mass of any body, the Newtonian constant of gravitation and the velocity of light. Show that in this system mass, length and time are all measured in terms of the same unit. If this unit is chosen to be the centimetre, what is the measure in centimetres of (*a*) a time of 1 s, (*b*) a mass of 1 g?

100. In the electromagnetic system, unit current is defined in terms of the force exerted by one current-carrying conductor on another, the permeability of free space being defined as unity; the unit of charge is derived from the unit of current. In the electrostatic system, on the other hand, the unit of charge is defined in terms of the force between charges, the dielectric constant of free space being defined as unity. Show that these two units of charge are different, their ratio being a quantity having the dimensions of velocity.

This velocity is about 3×10^8 m s^{-1}, and is the velocity of light in free space.

In the rationalized M.K.S. system, the unit of charge is one-tenth of the electromagnetic unit, and the force between two charges Q_1 and Q_2, a distance r apart in free space, is written

$$F = \frac{Q_1 Q_2}{4\pi\varepsilon_0 r^2}$$

Show that $\varepsilon_0 \approx 9 \times 10^{-12}$ in this system.

101. By dimensional arguments show that the speed of waves in a deep body of liquid is independent of the liquid density if the waves are long enough to be controlled by gravity, but not if they are so short as to be controlled by surface tension. Show also that for the former it varies as $\lambda^{\frac{1}{2}}$, for the latter as $\lambda^{-\frac{1}{2}}$, λ being the wavelength.

102. A number of clean glass tubes of different internal diameter are dipped into water and withdrawn, when it is found that all tubes with diameters less than 8.0 mm retain a drop of liquid filling the bore. When the same experiment is performed with chloroform, only tubes of diameter less than 4.0 mm retain a drop of liquid. If the surface tension of water is 72 mN m^{-1} and the specific gravity of chloroform 1.50, what is the surface tension of chloroform?

103. Which is in principle the speedier method of arriving at the antipodes: (*a*) free fall in a hole bored through the earth, or (*b*) travel in an earth satellite?

104. A spherical planet of radius R has a uniform density ρ and does not rotate. Assuming that the planet is all liquid show that the pressure at any point distant r from the centre is

$$\frac{2\pi\rho^2}{3}(R^2 - r^2)G.$$

Taking ρ to be 5 Mg m^{-3} estimate the pressure at the centre of the earth.

105. The value of g at the bottom of a mine-shaft 500 m deep is found to be 0.005% greater than at the top. Taking the mean density of the earth to be 5 Mg m^{-3} and assuming spherical symmetry, calculate the mean density of the earth's crust down to a depth of 500 m.

106. A mass of 10 kg is hung by a long wire from a balance arm and counterpoised. A large flat sheet of lead, 0.2 m thick, is then brought under the mass. What weight is needed on the other side of the balance to restore equipoise, if the influence of the lead on all but the 10 kg mass is negligible? [Density of lead = 11.4 Mg m^{-3}.]

107*. Explain why there are two tides a day. Show that if the earth were covered by an ocean of uniform depth, the range of the tide would be $3mR^4/(2MD^3)$ (i.e. about 0.5 m), where m = mass of moon, M = mass of earth, R = radius of earth, D = distance between moon and earth. Show also that tidal friction causes the moon to recede.

108. A sphere of volume v and density ρ_1 is suspended, by means of a thread attached to an independent support, below the surface of an incompressible liquid of density ρ_2 contained in a tall jar; the thread is cut and the sphere falls, reaching its terminal velocity before it hits the bottom. If the weight of jar and liquid together is W, what is the apparent weight of the system (i.e. the reaction between the jar and the surface on which it stands) (*a*) before the thread is cut; (*b*)* immediately after the thread is cut; (*c*) when the sphere has reached its terminal velocity?

[*Note.* An exact solution of (*b*) is difficult, but the principles involved should be discussed.]

109. A long uniform beam of square cross-section floats freely with a fraction f not submerged. Show that the stable position of the beam is:

(*a*) with two faces horizontal if $f \ll 1$;

(*b*) with two edges touching the surface if $f = \frac{1}{2}$.

* Show further that for values of f between $\frac{1}{2}(1-[1/\sqrt{3}])$ and $\frac{9}{32}$ the stable position is unsymmetrical, and that in particular for $f = \frac{1}{4}$ the beam floats with one edge touching the surface.

110. A Venturi meter in a water pipe consists of a slow taper from 100 mm diameter to 80 mm diameter. It is observed that a pressure difference equivalent to 150 mm head of water exists between the ends of the taper. Estimate the water flow in $m^3 s^{-1}$.

111. A uniformly spinning bowl of mercury has been used as a parabolic reflector for light. What speed of rotation is needed for a focal length of 1 m?

112. Oil of viscosity 0.6 kg $m^{-1} s^{-1}$ and density 0.9 Mg m^{-3} flows downhill in a shallow flat channel 1 m wide, whose slope is 1 in 1000. If the oil is everywhere 50 mm deep, calculate the volume flowing per second and the velocity of flow at the surface. Neglect edge effects.

113. The two arms of a U-tube of diameter 10 mm are connected by a capillary of radius 0.5 mm and length 0.1 m. Initially water stands in the U-tube with a difference of level of 0.2 m. How long will it take for the difference to become 20 mm, if the viscosity of water at the temperature of the experiment is 1 g m^{-1} s^{-1}.

114. A gas container has a volume of 1 m^3. The wall of the container is 10 mm thick, and there is a crack in it 1 μm wide and 0.1 m long. If the container is filled with air at a pressure 2 atm above atmospheric pressure, how long will it take for the pressure to fall to 1 atm above atmospheric? The temperature and pressure outside the container and the temperature inside remain constant throughout.

[Viscosity of air = 20 mg m^{-1} s^{-1}.]

115. A tube of circular bore has a slow conical taper from radius $2r$ at one end to radius r at the other. What would be the radius of a uniform tube of the same length which presented the same resistance to streamline flow?

116. A drop of oil of density 0.92 Mg m^{-3} falls freely under gravity in air of density 1.2 kg m^{-3} and viscosity 18.3 mg m^{-1} s^{-1} and attains a terminal velocity of 0.86 mm s^{-1}. When a vertical electric field of 30 V mm^{-1} is applied the downward velocity of the drop changes to 0.81 mm s^{-1}. Assuming the validity of Stokes's Law calculate the charge on the drop.

117*. In a number of experiments on the terminal velocity of spheres falling through viscous fluids the following results were obtained.

(*a*) Aluminium spheres (density = 2.7 Mg m^{-3} in isopropyl alcohol (density = 0.80 Mg m^{-3}, viscosity = 4.50 g m^{-1} s^{-1}):

Diameter of sphere	1.5	3.0	6.0	12.0 mm
Terminal velocity	0.167	0.33	0.58	0.88 m s^{-1}

(*b*) Steel spheres (density = 7.83 Mg m^{-3}) in olive oil (density = 0.93 Mg m^{-3}, viscosity = 99 g m^{-1} s^{-1}):

Diameter of sphere	10.0	17.5	30.0	52.5 mm
Terminal velocity	0.89	1.50	2.65	3.30 m s^{-1}

Examine these results critically in the light of dimensional analysis, and deduce a value for the terminal velocity of a spherical hailstone (density = 0.90 Mg m^{-3}) of diameter 2 mm in air (density = 1.3 kg m^{-3}; viscosity = 17 mg m^{-1} s^{-1}).

[Note that in these experiments Stokes's Law was not obeyed.]

118*. A flat plate surrounded by a liquid of viscosity η and density ρ oscillates sinusoidally in its own plane with angular frequency ω. Show that the effective inertia of the plate is increased as though it dragged with it on each side a layer of liquid of thickness $(\eta/2\omega\rho)^{\frac{1}{2}}$.

119. A long circular rod of radius r and length l is immersed in a large vessel containing a liquid whose viscosity is η. Show that to keep the rod turning slowly about its axis with angular velocity ω a torque $4\pi\eta r^2\omega l$ must be applied continuously.

120. Calculate the extension under its own weight of a wire 10 m long suspended freely from its top end, if the velocity of sound along the wire is 1 km s^{-1}.

121. A horizontal steel bar is 2 m long and its cross-section is a square of side 10 mm. One end of the bar is hinged to a fixed support and another support is placed beneath the beam at its mid-point. Calculate the load which must be applied half-way between the supports to produce a deflexion of 10 mm at the free end.

[Young's modulus for steel = 2×10^{11} N m^{-2}.]

122. A horizontal beam of uniform rectangular section, 2×1, has one end built into a wall, a diagonal of the rectangle being vertical. When a weight is hung on the free end, in what direction will the beam deflect?

123. A straight tube 1 m long, of radius 10 mm and wall thickness 100 μm, is closed at both ends. It is found that when the pressure inside is increased from 1 to 10 atm the tube lengthens by 100 μm. What is the bulk modulus of the material of the tube?

124*. A straight vertical strut whose length is l and whose cross-section is a square of side a is fixed firmly to the ground. Show that the maximum load it can carry on its free end is $\pi^2a^4E/48l^2$, where E is Young's modulus for the material of the strut.

125. Among the types of sound wave that can be propagated through a solid are the following:

(*a*) Longitudinal compressional waves along a thin rod.

(*b*) Longitudinal compressional waves along a thin sheet, whose dimensions apart from thickness are much greater than the wavelength.

(*c*) Longitudinal compressional waves in an extended solid.

(*d*) Shear waves in an extended solid.

(*e*) Torsional waves along a thin rod.

Show that for an isotropic solid for which Poisson's ratio is σ these velocities are in the ratios

$$1:\left[\frac{1}{1-\sigma^2}\right]^{\frac{1}{2}}:\left[\frac{1-\sigma}{(1+\sigma)(1-2\sigma)}\right]^{\frac{1}{2}}:\left[\frac{1}{2(1+\sigma)}\right]^{\frac{1}{2}}:\left[\frac{1}{2(1+\sigma)}\right]^{\frac{1}{2}}.$$

126*. A long uniform bar is freely supported without tension and caused to transmit flexural waves by shaking one end sideways. Show that the lateral displacement y obeys the differential equation

$$\frac{\partial^4 y}{\partial x^4}+C\frac{\partial^2 y}{\partial t^2}=0,$$

in which C is a constant determined by the characteristics of the bar. If a sine wave of frequency 10 Hz has wavelength 0.5 m what is the wave-length of a 20 Hz wave? [Assume that each element of the bar vibrates sideways without rotation.]

127*. When a very viscous liquid is suddenly sheared by a small angle θ the tangential shear stress S rises suddenly to a value $n\theta$ (n is the 'instantaneous shear modulus') and then decays exponentially with time constant τ.

(*a*) Show that this behaviour is described by the equation

$$\frac{\mathrm{d}S}{\mathrm{d}t}+\frac{S}{\tau}=n\frac{\mathrm{d}\theta}{\mathrm{d}t}.$$

(*b*) If S oscillates with angular frequency ω, show that the corresponding oscillations of θ lag by a phase angle $\cot^{-1}(\omega\tau)$.

(*c*) Show that the equation of transverse wave motion takes the form

$$\frac{\partial^2 y}{\partial t^2}+\frac{1}{\tau}\frac{\partial y}{\partial t}=\frac{n}{\rho}\frac{\partial^2 y}{\partial x^2},$$

where ρ is the density, and that this equation has solutions describing exponentially damped waves.

(*d*) If the velocity of high frequency ($\omega\tau \gg 1$) transverse waves is 1 km s^{-1}, the viscosity of the liquid 100 Mg m^{-1} s^{-1}, and its density 2 Mg m^{-3}, what is the value of τ, and how far does the wave travel before its amplitude is reduced by a factor e?

128. A glass capillary tube is held vertical with its lower end in a liquid of density ρ and surface tension γ which wets the glass completely. The tube has a slight taper so that the radius is $r = r_0 - \alpha h$, where h is the height above the surface of the liquid. Show that if the constant $\alpha < g\rho r_0^2/8\gamma$ there are two heights for which the meniscus will be in equilibrium, and find them. For which of them would you expect equilibrium to be stable? What would you expect to happen if $\alpha > g\rho r_0^2/8\gamma$?

129. Two equal soap bubbles are formed on the two ends of a cylindrical tube. Both are larger than one hemisphere. Show that the equilibrium of this system is unstable, and describe what happens when a small disturbance makes the radius of one bubble change slightly. What would happen if both bubbles were initially smaller than one hemisphere?

130. Show that the vapour pressure in equilibrium with a drop of radius r is greater than that in equilibrium with a flat liquid surface by an amount approximately equal to $2\gamma\rho_G/(r\rho_l)$, where γ is the surface tension and ρ_G and ρ_l the vapour and liquid densities.

The air in a Wilson cloud chamber is initially saturated with water vapour at 15.0 °C, and the volume is then increased adiabatically by 5%. If condensation does not occur, calculate the resulting temperature, the degree of supersaturation (i.e. ratio of vapour pressure to equilibrium vapour pressure), and the radius of an electrically neutral droplet just large enough to continue growing in the supersaturated vapour.

[Take $c_p/c_V = 1.4$ and the vapour pressure of water from the table in question 185; $\gamma = 70$ mN m^{-1}.]

131*. In a careful series of experiments Harkins and Brown measured the mass of liquid drops falling from the end of a pipette. In the results quoted below, ρ denotes the density of the liquid in Mg m^{-3} and γ its surface tension in mN m^{-1}. For water ($\rho = 0.998$, $\gamma = 72.80$)

Radius of pipette/mm	Mass of drop/mg
2.305	68.0
3.502	98.7
3.997	111.6

For benzene ($\rho = 0.880$, $\gamma = 28.88$)

Radius	Mass
1.972	22.41
2.305	25.78
2.680	29.74

In another experiment it is found that drops of ether ($\rho = 0.714$) falling from a pipette of radius 1.800 mm have mass 11.95 mg. Assuming gravity to be the same in all experiments, determine by dimensional analysis the surface tension of ether.

132. If the temperature at which water has a maximum density (T_{max}) were determined using a water-in-glass dilatometer, how much would the result be in error? The coefficient of cubical expansion of glass is 2.3×10^{-5} K^{-1}; the density of water at temperature T, ρ_T, obeys the equation $\rho_{max} - \rho_T = 7.5(T - T_{max})^2$g m^{-3}.

133. A bimetallic strip consists of two tapes $\frac{1}{2}$ mm thick, one of stainless steel whose linear expansion coefficient α is 10^{-5} K^{-1} and the other german silver for which $\alpha = 1.8 \times 10^{-5}$ K^{-1}; the tapes are welded face-to-face. A 0.1 m length of the strip is firmly fixed at one end to a support. How much does the other end move when the temperature is changed by 100 K?

134. A cylindrical gas jar 0.2 m deep is full to the brim with water at 300 K, and is covered with a porous membrane, so that a current of dry air blowing over the membrane carries away all water molecules reaching the top of the jar, but does not disturb the air inside. How long will the jar take to empty completely?

[Density of saturated water vapour at 300 K = 30 g m^{-3}; diffusion coefficient of water vapour through air = 30 mm^2 s^{-1}.]

135. A spherical vessel of radius 0.25 m containing liquid at 120 °C is to be thermally insulated by covering it with material of thermal conductivity 0.6 W m^{-1} K^{-1}. The rate of loss of heat per unit area from the outside of the insulation is 3θ W m^{-2}, where θ is the temperature difference between it and the surroundings. If the surroundings are at 20 °C, show that the heat lost per second from the vessel is increased by increasing the thickness of the insulation until a certain critical thickness is reached, after which the heat lost per second decreases. Find the heat lost per second when the insulation is (*a*) very thin, (*b*) very thick, (*c*) of the critical thickness referred to above.

136. A pond is covered with a layer of ice 50 mm thick. How long will it be before the ice is 0.1 m thick if the air temperature stays constant at −10 °C? Assume the following values for the properties of ice: density = 0.9 Mg m^{-3}; thermal conductivity = 2.1 W m^{-1} K^{-1}, latent heat of fusion = 330 J g^{-1}.

137*. A thick plane slab, whose thermal conductivity is 10 W m^{-1} k^{-1} and whose thermal capacity per unit volume is 4 J cm^{-3} k^{-1}, is subected on its surface to a regular succession of short pulses of radiant energy, one every 5 s. Show that at a sufficient depth below the surface the temperature oscillations are approximately sinusoidal, and estimate the necessary depth.

138. The tungsten filament of an electric light bulb is 50 μm in diameter and runs at a temperature of approximately 3 kK when operated on the d.c. mains. When it is operated at the same power rating on the 50 Hz a.c. mains the intensity of its light output oscillates about its mean value by approximately 25%. Estimate the order of magnitude of Stefan's constant.

[Heat capacity of tungsten per unit volume = 4.2 MJ m^{-3} K^{-1}.]

139. A large plane sheet of emissivity ε_1 and temperature T_1 faces another sheet of emissivity ε_2 and temperature T_2. Show that the rate of exchange of heat between the two due to radiation is

$$\sigma(T_2^4 - T_1^4)\varepsilon_2\varepsilon_1/(\varepsilon_1 + \varepsilon_2 - \varepsilon_1\varepsilon_2)$$

per unit area, where σ is Stefan's constant.

140*. A large storage vessel for liquid oxygen may be considered as a perfectly evacuated spherical Dewar vessel of inner radius 1 m and outer radius 1.2 m. Treating the walls of the vessel as perfectly black, calculate the rate of loss of liquid oxygen due to radiation. How will the rate of loss be changed by the interposition of an insulated spherical copper radiation shield midway between the inner and outer walls?

[Reflexion coefficient of copper = 0.98 assumed independent of temperature; room temperature = 300 K; temperature of liquid oxygen = 90 K; latent heat of oxygen = 240 J g^{-1}.]

141. A compressor takes in air at 300 K and 1 atm pressure and delivers compressed air at 2 atm, consuming 200 W of useful power. If the compression is reversible and adiabatic, what volume will it deliver per second and at what temperature?

[For air $\gamma = 1.4$.]

142. A vessel of 10^{-2} m^3 capacity has an accurately ground vertical tube of diameter 10 mm and length 1 m connected to it, and is filled with gas at a pressure of 10^5 N m^{-2}. A well-fitting steel ball-bearing (density 8 Mg m^{-3}) is allowed to fall from rest down the tube, and is found to oscillate at a frequency of 0.79 Hz. What is the value of γ for the gas, and what is the initial amplitude of the oscillation if damping is small?

143. A mixture of 0.1 mol of helium ($\gamma = \frac{5}{3}$) with 0.2 mol of nitrogen ($\gamma = \frac{7}{5}$) is initially at 300 K and occupies 4×10^{-3} m^3. Show that the changes of temperature and pressure of the system which occur when the gas is compressed slowly and adiabatically can be described in terms of some intermediate value of γ. Calculate the magnitude of these changes when the volume is reduced by 1%.

144. Two thermally insulated cylinders, *A* and *B*, of equal volume, both equipped with pistons, are connected by a valve. Initially *A* has its piston fully withdrawn and contains a perfect monatomic gas at temperature *T*, while *B* has its piston fully inserted, and the valve is closed. Calculate the final temperature of the gas after the following operations, which each start with the same initial arrangement. The thermal capacity of the cylinders is to be ignored.

(*a*) The valve is fully opened and the gas slowly drawn into *B* by pulling out piston *B*; piston *A* remains stationary.

(*b*) Piston *B* is fully withdrawn and the valve is opened slightly; the gas is then driven as far as it will go into *B* by pushing home piston *A* at such a rate that the pressure in *A* stays constant; the cylinders are in thermal contact.

(*c*) As in (*b*), but the cylinders remain thermally isolated (calculate the temperature in each cylinder).

(*d*) Piston *B* is fully withdrawn and the valve opened; the two cylinders are placed in thermal contact.

(*e*) As in (*d*), but the cylinders remain thermally isolated (calculate the temperature in each cylinder).

145*. A differential gas thermometer for use at very low temperature consists of two identical bulbs, each of capacity 1 cm^3, connected by fine capillary tubes to opposite sides of a manometer at room temperature (300 K), containing oil of density 0.8 Mg m^{-3}. The manometer tubing is 3 mm in diameter, and there is a dead space of 1 cm^3 above each meniscus. If the system is filled with helium gas at a pressure of 100 mmHg when the bulbs are at 3 K, what level difference is produced in the manometer by a temperature difference of 10 mK between the bulbs? Take helium to be a perfect gas and neglect the dead space in the capillary tubes.

146. Heat is released at the earth's surface at the rate of about 80 kJ m^{-2} per minute. Estimate the temperature gradient in the earth's atmosphere on the assumptions: (*a*) that no convection currents occur and that all the heat is conducted upwards, the thermal conductivity of air being 25 mW m^{-1} k^{-1}, (*b*) that convection does occur and that the convected gas expands adiabatically as it rises. Which of the two is more likely to occur in practice?

147. Show that for a gas obeying van der Waal's equation, $(p+a/V^2)(V-b) = RT$, the Boyle temperature is 27/8 times the critical temperature. The critical pressure of a certain gas is 40 atm; up to what pressure will the product pV change by less than 0.1% at the Boyle temperature?

148. Show that for a gas obeying Dieterici's equation of state, $p(V-b) = RTe^{-a/RTV}$, the ratio $pV/(RT)$ is equal to 0.27 at the critical point.

149. A sealed vessel of 50 cm^3 capacity contains at room temperature 10 cm^3 of water, the remaining space containing only water vapour. Discuss what will happen as the vessel is heated, and draw rough curves showing the variation with temperature of (*a*) the pressure in the vessel, (*b*) the thermal capacity of the system as a whole. Values of the density, ρ, of saturated water vapour at different temperatures are given below:

T/°C	350	360	370	380
ρ/Mg m^{-3}	0.11	0.15	0.21	0.37

150. The vapour pressure of neon varies with absolute temperature according to the equation

$$\log_{10}p = 8.75-0.0437T-127T^{-1}$$

(p being measured in mm of mercury). Assuming the law of corresponding states, estimate the vapour pressure of argon at 135 K. [The critical pressure for neon is 27 atm and its critical temperature is 44 K: the corresponding quantities for argon are 48 atm and 151 K.]

151. Two constant-volume gas thermometers contain different gases A and B whose critical temperatures are 500 and 400 K respectively.

The volume of the thermometer is the same fraction of the critical volume for each gas. The temperature is taken to be exactly proportional to the pressure, the ice-point being fixed at 273.16 K, i.e. the same as on the absolute scale. With this procedure thermometer A reads as follows:

Absolute temperature	300	320	340	360
Measured temperature	301.2	322.5	344.1	366.5
Absolute temperature	380	400	420	
Measured temperature	389.6	413.3	438.0	

If the gases obey the law of corresponding states, what will thermometer B read when the absolute temperature is 320 K?

152. Two similar, wide, uniform glass tubes, closed at one end, are inverted over mercury. One contains pure dry air, the other air and saturated water vapour, and in both the mercury levels inside and out are the same when the tops of the tubes are 1 m above the level of the mercury. Both tubes are now depressed until the tops are 0.5 m above the outside level of the mercury. In the first tube the level of the mercury inside is 253 mm below the level outside, and in the second 250 mm. Calculate the atmospheric pressure and the saturated pressure of the water vapour.

153. The usual way of cooling liquid helium contained in a Dewar flask is to cause it to boil under reduced pressure by pumping away the vapour. The latent heat of evaporation of liquid helium is 23 J g^{-1}, and the specific heat capacity of the liquid is given in the table below. What fraction of the liquid must be lost in reducing the temperature from the normal boiling point of 4.2 to 3.0 K?

T/K	4.5	4.0	3.5	3.0	2.5
c/Jg^{-1} K^{-1}	4.9	3.8	3.0	2.5	2.2

154. At what temperature would the R.M.S. velocity of nitrogen molecules in the earth's atmosphere equal the minimum velocity for escape from the earth's gravitational field?

155. Show that the rms velocity of the molecules in a gas is $(3/\gamma)^{\frac{1}{2}}$ times the velocity of sound.

156. Show that for an ideal gas the pressure p is $\frac{2}{3}$ times the translational kinetic energy of the gas per unit volume, and that this is consistent with the laws: (*a*) pV = constant for an isothermal expansion; (*b*) $pV^{5/3}$ = constant for a slow adiabatic expansion of a monatomic gas, during which the gas molecules lose kinetic energy by colliding elastically with a moving piston. By considering the sharing of energy show that if the gas molecules have n degrees of freedom in addition to their translational motion pV^{γ} = constant in an adiabatic expansion, γ being $(5+n)/(3+n)$.

157. Show that the number of impacts of gas molecules on unit area in unit time is $\frac{1}{4}n\bar{c}$, where n is the number of molecules per unit volume and $\bar{c}$ is their mean velocity.

(*a*) A clean tungsten filament is introduced into a vessel containing oxygen at a pressure of 10^{-4} mmHg and a temperature of 300 K. If every oxygen molecule which hits the filament remains attached, and if the molecules (regarded as spheres of diameter 0.3 nm) arrange themselves in a close-packed pattern, how long does it take for a monomolecular layer to be formed?

(*b*) A vessel partially filled with mercury (atomic weight 201), and closed except for a hole of area 0.1 mm^2 above the liquid level, is kept at 0 °C in a continuously evacuated enclosure. After 30 days it is found that 24 mg of mercury have been lost. What is the vapour pressure of mercury at 0 °C?

[In a gas of molecular weight M, $\bar{c} = (8RT/(M\pi))^{\frac{1}{2}}$.]

(*c*) A disc of radius r spins at an angular velocity ω in a vessel containing gas at so low a pressure that the free path is much greater than r. If molecules striking the disc are momentarily attached, and evaporated later in a random direction, show that the torque exerted by the gas on the disc is $\frac{1}{4}\pi\rho\bar{c}\omega r^4$, ρ being the density of the gas.

(*d*) Two equal vessels initially containing air at 0 °C and a pressure of 10^{-2} mmHg are connected by a hole of diameter 0.1 mm, much less than the free path. If one vessel is now heated to 100 °C, what pressure difference is established between the vessels?

158*. According to the Maxwell–Boltzmann distribution law the fraction $f(c)\mathrm{d}c$ of molecules in a gas at temperature T whose velocity lies between c and $c+\mathrm{d}c$ is given by the expression

$$f(c) = Ac^2\mathrm{e}^{-mc\ /2kT},$$

in which m is the mass of the molecule, and k is Boltzmann's constant.

(*a*) Show that $A = (2/\pi)^{\frac{1}{2}}(m/kT)^{\frac{3}{2}}$ and calculate the mean velocity $\bar{c}$, the R.M.S. velocity, the most probable velocity, and the mean kinetic energy.

(*b*) A gas at low pressure and at temperature T is contained in a vessel from which it leaks through a small hole whose diameter is much less than the molecular mean free path. Show that the mean kinetic energy of the molecules leaving the vessel is $2kT$.

(*c*) In a Pirani gauge for measuring low pressures a thin wire is held at an elevated temperature by passing a current through it. If the diameter of the wire is much less than the free path the molecules may be assumed to hit the wire as from a gas at room temperature, and leave with the velocity distribution characteristic of a gas at the temperature of the wire. What heat per unit length is needed to maintain a filament of diameter 0.1 mm 100 K above room temperature in helium at a pressure of 10^{-2} mmHg if other sources of heat loss are neglected? At what gas pressure does the loss through the gas equal the radiation loss, if the wire is black?

(*d*) Show that the fraction of the molecules of a gas which hit a given surface with a kinetic energy greater than ε is $(1+[\varepsilon/kT])e^{-\varepsilon/kT}$. A molecule AB dissociates if it hits the surface of a solid catalyst with a kinetic energy greater than 0.6 eV. Show that the rate of the reaction $AB \rightarrow A+B$ is doubled by raising the temperature about 10 K from room temperature.

$$\left[\text{If } I_n = \int_0^\infty x^n e^{-x^2} dx, \text{ then } I_n = \tfrac{1}{2}(n-1)I_{n-2},\ I_1 = \tfrac{1}{2},\ I_0 = \tfrac{1}{2}\sqrt{\pi}.\right]$$

159. If the molecules of methane (molecular weight 16) and argon (atomic weight 40) are regarded as spheres, and given that the densities of the liquids are 0.42 and 1.41 Mg m^{-3} and that $c_p/c_V = 1.29$ and 1.67 respectively, estimate

(*a*) the ratio of volumes of the molecules;

(*b*) the ratio of viscosities and thermal conductivities at N.T.P. [The measured values are 0.5 and 1.7.]

160. At 0 °C and a pressure of 10^5 N m^{-2} the viscosity of nitrogen (molecular weight 28) is 16.7 mg m^{-1} s^{-1} and the density is 1.25 kg m^{-3}. The density of liquid nitrogen is 0.78 Mg m^{-3}. Use this information to estimate the diameter of the nitrogen molecule and Avogadro's number.

161. The thermal conductivity of helium at N.T.P. is 0.14 W m^{-1} K^{-1}. At what pressure at 0 °C is the mean free path 10 mm long?

162. What values would you expect for γ, the ratio of specific heat capacities at constant pressure and constant volume, for the following gases, if equipartition of energy prevailed?—

(*a*) helium (monatomic);
(*b*) hydrogen (diatomic);
(*c*) carbon dioxide (OCO collinear);
(*d*) water vapour (HOH not collinear);
(*e*) benzene.

Rotation of individual atoms may be ignored.

163. The potential energy V between the two atoms in a hydrogen molecule is given by the following (empirical) expression, where r is the distance between the atoms and D, a and r_0 are constants with the values given below:

$$V = D\{e^{-2a(r-r_0)} - 2e^{-a(r-r_0)}\}.$$

Calculate the quantum of energy required to set the molecule:

(*a*) rotating (this may be taken to be $\hbar^2/2I$, where $\hbar$ is Planck's constant and I is the moment of inertia of the molecule);

(*b*) vibrating (this is $\hbar\omega$, where ω is the angular frequency of vibration).

Hence estimate the temperatures at which rotation and vibration begin to contribute to the specific heat of hydrogen gas.

[$D = 7 \times 10^{-19}$ J; $a = 2 \times 10^{10}$ m^{-1}; $r_0 = 8 \times 10^{-11}$ m.]

164*. In observations on the sedimentation equilibrium of gamboge particles in water at a temperature of 288 K, Perrin obtained the following results:

Height of layer/μm	5	35	65	95
Mean number of particles in layer	100	47	22.6	12

When a substantial number of the particles were released at the top of a jar of water he found that they settled out with a mean velocity of 61 μm per hour. On the assumption that Stokes's Law is obeyed, deduce a value for Boltzmann's constant k.

[Density of gamboge = 1.19 Mg m^{-3}; viscosity of water = 1.13 g m^{-1} s^{-1}.]

165. Liquid argon consists of spherical atoms of diameter d packed fairly closely together. Any pair of atoms in contact attract one another, so that it takes an amount of energy ε to separate them. Obtain an approximate expression for the energy U required to separate all the atoms in unit volume of the liquid.

Rough estimates of U may be obtained from

(*a*) the latent heat of evaporation per unit volume;

(*b*) the work required to draw out unit volume into a film only one atom thick against the forces of surface tension. Use the following data to estimate both d and ε.

Density of liquid = 1.4 Mg m^{-3}; latent heat of evaporation = 160 J g^{-1}; surface tension = 13 mN m^{-1}.

166. In the sodium chloride crystal the Na^+ and Cl^- ions are arranged alternately on a simple cubic lattice. The ions interact by ordinary electrostatic forces, and they also repel one another in such a way that any two ions a distance x apart have a positive potential energy proportional to $1/x^n$. Show that the bulk modulus of elasticity K should be given by $nU/9V$, where U is the energy required to separate completely all the ions in volume V of the crystalline salt. From the data below obtain a value for n.

Molecular weight of NaCl = 58.5, density = 2.17 Mg m^{-3}; K = 2.4×10^{10} N m^{-2}; U = 760 kJ mol^{-1}.

167. According to Einstein, due to quantum effects the specific heat c_V of a solid falls at low temperatures from the value $3R$ expected if equipartition of energy holds, and is only $\frac{3}{2}R$ at a temperature T at which $h\nu \doteqdot 3kT$, ν being the frequency of oscillation of an atom in the solid. According to Lindemann, a solid melts when the amplitude of atomic vibrations reaches a certain fraction of the interatomic spacing.

Use these ideas, and the following data for copper and lead, which both have face-centred cubic structures, to estimate the melting-point of lead.

Copper. $c_V = \frac{3}{2}R$ at 81 K; atomic weight = 63.5; density = 9.0 Mg m^{-3}; melting point = 1356 K.

Lead: $c_V = \frac{3}{2}R$ at 22 K; atomic weight = 207; density = 11.3 Mg m^{-3}.

168. A perfect gas, for which $\gamma = 1.50$, is used as the working substance in a cylinder, undergoing a Carnot cycle. At the source temperature (240 °C) it is expanded from a pressure of 10 atm and a volume of 10^{-3} m^3 to a pressure of 4 atm and a volume of 2.5×10^{-3} m^3.

(*a*) Between what limits of pressure and volume does it operate at the sink temperature (50 °C)?

(*b*) Calculate the heat taken from the source and the heat rejected to the sink, in each cycle, and the efficiency, assuming perfect reversibility.

(*c*) If the area of the piston is 0.02 m^2, and a force of 100 N is needed to overcome friction between it and the cylinder, calculate the reduction in efficiency due to friction.

169. For a certain perfect gas $c_p = 29.4$ and $c_V = 20.9$ J k^{-1} mol^{-1}. In a reversible heat engine the gas is (*a*) heated at constant volume until the pressure is $\frac{6}{5}$ of its initial value, (*b*) heated at constant pressure until the volume is $\frac{5}{4}$ of its initial value, (*c*) cooled at constant volume until the pressure returns to its initial value, (*d*) cooled at constant pressure until the volume returns to its initial value. What is the ratio of the efficiency of this engine to that of a Carnot engine working between the same extremes of temperature?

170. Calculate the maximum useful work obtainable from the following systems of source and sink:

	Source	Sink
(*a*)	10^3 m^3 of water at 100 °C and 1 atm	A lake at 10 °C
(*b*)	10^3 m^3 of water at 100 °C and 1 atm	10^3 m^3 of water at 10 °C and 1 atm
(*c*)	10^6 kg of steam at 100 °C and 1 atm	A lake at 10 °C
(*d*)	10^3 m^3 of a perfect monatomic gas at 10 °C and 10 atm	A lake at 10 °C
(*e*)	10^3 m^3 of a perfect monatomic gas at 100 °C and 10 atm	A lake at 10 °C

[Specific heat capacity of water $= 4.2\ \mathrm{J\ g^{-1}\ K^{-1}}$, latent heat of vaporization $= 2.3\ \mathrm{kJ\ g^{-1}}$.]

171. Calculate the changes in the total entropy of the universe as a result of the following processes:

(*a*) A copper block, of mass 400 g and thermal capacity $150\ \mathrm{J\ K^{-1}}$ at 100 °C is placed in a lake at 10 °C.

(*b*) The same block at 10 °C is dropped from a height of 100 m into the lake.

(*c*) Two such blocks, at 100 and 0 °C, are joined together.

(*d*) A 1 μF capacitor is connected to a 100 V reversible battery at 0 °C.

(*e*) The same capacitor, after charging to 100 V, is discharged through a resistor at 0 °C.

(*f*) One mole of a perfect gas at 0 °C is expanded reversibly and isothermally to twice its initial volume.

(*g*) The same expansion as in (*j*) is carried out reversibly and adiabatically.

(*h*) The same expansion as in (*f*) is carried out by opening a valve to an evacuated container of equal volume.

172*. Three identical bodies of constant thermal capacity are at temperatures 300, 300 and 100 K. If no work or heat is supplied from outside, what is the highest temperature to which any one of the bodies can be raised by the operation of heat engines?

173. A perfect gas is defined as having the following properties:

(i) The Joule coefficient vanishes, i.e. there is no temperature change accompanying expansion into a vacuum.

(ii) It obeys Boyle's Law, i.e. at a constant temperature pV is constant.

Show that it follows from this definition that for a perfect gas:

(*a*) The internal energy U and the specific heat capacities c_p and c_V are functions of temperature only.

(*b*) $pV = RT$, where T is the absolute temperature and R a constant.

(*c*) $c_p - c_V = R$

(*d*) There is no temperature change accompanying Joule–Kelvin expansion.

174. A gas obeys the equation

$$p(V-b) = RT,$$

and has c_V independent of temperature. Show that:

(*a*) The internal energy is a function of temperature only.

(*b*) The ratio $\gamma(= c_p/c_V)$ is independent of temperature and pressure.

(*c*) The equation of an adiabatic change has the form

$$p(V-b)^\gamma = \text{constant}.$$

175*. Show that for a gas obeying van der Waals's equation,

$(p+a/V^2)(V-b) = RT$, and having c_V independent of temperature, $T(V-b)^{R/c_V}$ is constant in an adiabatic expansion.

176. In a certain compressor a perfect gas at room temperature T_0 and atmospheric pressure p_0 is compressed adiabatically, and is then passed through water-cooled tubes until eventually it emerges at pressure p_1 and temperature T_0. Find an expression for the work required for this process, compared with what would be needed for a reversible isothermal compression leading to the same result, and show that the ratio is not less than unity. Discuss also the changes of entropy occurring in the two processes.

[Note that if $a > 1$ and $x > 0$, $a^x > 1+x\ln a$.]

177*. A Simon helium liquefier consists essentially of a vessel into which helium gas is compressed to a high pressure p at 10 K (above the critical point of helium). The vessel is then thermally isolated, and the helium is allowed to escape slowly through a capillary tube until the pressure within the vessel is 1 atm, and the temperature 4.2 K, the normal boiling-point of helium. Assuming that the thermal isolation is perfect, that the heat capacity of the vessel is negligible in comparison with that of the gas, and that gaseous helium obeys the perfect gas law, calculate what value of p must be chosen for the vessel to be entirely filled with liquid.

[Latent heat of liquid helium at 4.2 K = 84 J mol^{-1}; c_V of gaseous helium = 12.6 J K^{-1} mol^{-1}.]

178. To a thermally isolated vessel are connected a number of tubes through which fluid is led into or out of the vessel. Show that when the flow has settled down to a stationary condition, the enthalpy content of the fluid entering is the same as that leaving, provided there is no heat conduction along the tubes at the points of attachment.

A helium liquefier in its final stage of liquefaction takes in compressed helium gas at 14 K, liquefies a fraction α and rejects the rest as gas at 14 K and atmospheric pressure. Use the data below to determine the input pressure which allows α to take its maximum value, and determine what this value is.

Enthalpy of liquid helium at atmospheric pressure = 10.1 J g^{-1}.

Pressure	0	10	20	30	40 atm
Enthalpy of helium gas at 14 K	87.4	78.5	73.1	71.8	72.6 J g^{-1}

179*. Show that in a Joule-Thomson expansion no temperature change occurs if $(\partial V/\partial T)_p = V/T$. Hence calculate the ratio of inversion temperature to critical temperature for a gas whose equation of state is

$(a)\ (p+a/V^2)(V-b) = RT,$

$(b)\ p(V-b) = RT\exp[-a/(RTV)].$

180. The equation of state for helium gas may be expressed as a virial series

$$pV = RT(1+B/V+\ldots),$$

in which B is a function of temperature only. The following table gives some values of B for 1 mol of helium:

T/K	B/cm³	T/K	B/cm³
10	−23.3	50	7.6
20	−4.0	60	8.9
30	+2.4	70	9.8
40	5.6		

Determine the Boyle temperature and the Inversion temperature.

181. A vessel containing helium gas at 26 K and 1 atm pressure is placed within a much larger evacuated vessel which is thermally isolated. The inner vessel is then punctured. Using the data of the preceding question and taking $c_V = \frac{3}{2}R$, calculate the resulting drop in temperature of the gas.

182*. An elastic filament is such that when stretched by a force F at a temperature T, the extension x is given by the equation

$$\mu x = \alpha t + F,$$

where $\mu = \mu_0(1+\beta t)$ and $t = T-T_0$; T_0 is a fixed temperature, and α, β and μ_0 are positive constants. When the filament is maintained at a constant length and heated, its thermal capacity is found to be proportional to the temperature: $C_x = AT$. Show that:

(*a*) A is independent of x.

(*b*) If the entropy is S_0 when $t = 0$ and $x = 0$, then

$$S = S_0 + \alpha x - \tfrac{1}{2}\mu_0\,\beta x^2 + At.$$

(*c*) If the filament is heated under no tension the thermal capacity $C_F = (A+[\mu_0^2\alpha^2/\mu^3])T$.

(*d*) For small extensions under adiabatic conditions the filament cools and the appropriate elastic modulus is $\mu+\mu_0^2\alpha^2/(\mu^2 A)$.

(*e*) When the adiabatic extension is increased so that $x > \alpha/\beta\mu_0$ the filament starts to get warmer.

183. The surface tension of water changes linearly with temperature from 75 mN m^{-1} at 5 °C to 70 mN m^{-1} at 35 °C. Calculate the surface contributions to the internal energy and entropy.

An atomizer produces water droplets of diameter 0.1 μm. A cloud of droplets at 35 °C coalesces to form a single drop of water of mass 1 g:

estimate the temperature of the drop if there is no heat exchange with the surroundings. What is the increase of entropy in this process?

184. When lead is melted at atmospheric pressure the melting point is 327.0 °C, the density decreases from 11.01 to 10.65 Mg m^{-3} and the latent heat of fusion is 24.5 J g^{-1}. What is the melting point at a pressure of 100 atm?

185*. Show that the latent heat of evaporation of a liquid obeys the following equations:

$$L = RT^2 \frac{d}{dT}(\ln p),$$

$$\frac{dL}{dT} = (c_p)_{\text{vapour}} - (c_p)_{\text{liquid}},$$

provided that the vapour pressure p is small and the expansion coefficient of the liquid may be taken as zero. Examine the following experimental data for water in the light of these equations:

T/°C	p/mmHg	T/°C	p/mmHg
0	4.581	25	23.73
5	6.536	30	31.79
10	9.198	35	42.14
15	12.77	40	55.29
20	17.51		

c_p for water vapour = 1.9 J g^{-1} K^{-1}

186*. When a metal is heated *in vacuo* it emits electrons which form a gaseous phase in equilibrium with the 'condensed' electrons in the metal. The condensed electrons occupy negligible volume and have a negligible specific heat capacity; the gaseous electrons may be treated as a perfect gas with thermal energy $\frac{3}{2}kT$ per electron. A certain minimum amount of energy ϕ (the work function) is needed to remove an electron from the metal to a state of rest outside. By treating this system as a problem of phase equilibrium, show that the pressure p exerted by the electron gas in equilibrium varies with temperature according to the equation

$$p \propto T^{5/2} e^{-\phi/kT}.$$

By considering the kinetics of the equilibrium, on the assumption that any electron hitting the surface of the metal from outside has a fixed probability of entering the metal, show that the maximum current I that can be drawn from the metal by application of an electric field varies according to Richardson's Law

$$I \propto T^2 e^{-\phi/kT}.$$

The following table gives some experimental values of I for a tungsten filament. What is the value of ϕ in electron volts? What is the longest wavelength of light that will eject photo-electrons from the metal?

T/K	1900	2000	2100	2200	2250
I/mA	0.80	3.8	17	56	111

187. On the basis of the following information, which is partly hypothesis and partly somewhat simplified experimental data, calculate the melting pressure of the helium isotope ^{3}He at zero absolute temperature:

(*a*) Between 0 and 10 μK, the specific heat capacity of the solid is very high, but between 10 μK and 1 K it is much less than that of the liquid.

(*b*) The specific heat capacity of the liquid is proportional to T below 1 K.

(*c*) The expansion coefficient of both phases may be assumed to be zero.

(*d*) At 0.4 K the melting pressure p_m is 30 atm and $dp_m/dT = 0$; at 0.7 K, p_m is 33 atm.

188*. The high temperature behaviour of iron may be idealized as follows: below 900 °C and above 1400 °C α-iron is the stable modification, and between these temperatures γ-iron is stable. The specific heat capacity of each phase may be taken as constant, being 0.775 J g^{-1} K^{-1} for α-iron and 0.690 J g^{-1} K^{-1} for γ-iron. What is the latent heat at each transition?

189*. According to Debye's theory, the specific heat capacity of a crystalline solid may be expressed in the form

$$c_V = f(T/\Theta),$$

where Θ is independent of temperature but depends on volume according to the law $\Theta \propto V^{-\gamma}$, γ being a constant. Show that the expansion coefficient α and the isothermal compressibility κ obey the law

$$\alpha = \gamma c_V \kappa / V.$$

[Note that the third law of thermodynamics demands that α and c_V both vanish at zero absolute temperature.]

190. A thermocouple is made of two metals whose Thomson coefficients are each proportional to the absolute temperature; show that if one junction is kept at 0 °C and the other at t °C the thermoelectric voltage $u = at + bt^2$, where a and b are constants.

If u is expressed in μV, $a = 2.8$ and $b = 0.0060$ for a copper–lead couple, while $a = -38.1$ and $b = 0.045$ for a constantan–lead couple. Calculate the Peltier coefficient of a copper–constantan junction, and the difference between the Thomson coefficients of copper and constantan, at 100 °C. At what temperature does u for a copper–constantan couple have its maximum value?

191. By treating radiation in a cavity as a gas of photons whose energy E and momentum p are related by the expression $E = cp$, c being the velocity of light, show that the pressure exerted on the walls of the cavity is one-third of the energy density.

Show that when radiation contained in a vessel with perfectly reflecting walls is compressed adiabatically it obeys the equation $pV^{4/3}$ = constant.

192. In a cathode-ray tube, electrons from the cathode traverse a perpendicular deflecting field after falling through a potential difference of 1 kV. If the separation between the deflector plates is 10 mm, the effective length of the deflecting field 20 mm, and the fluorescent screen 0.2 m from the ends of the deflector plates, what is the length of the line traced on the screen when 100 V r.m.s. is applied to the plates?

193. Assuming that the earth's field may be represented as due to a short magnet at the earth's centre and that the horizontal component of the flux density at magnetic latitude 54° is 18 μT, calculate the density and direction of the flux at the earth's surface (*a*) at the magnetic poles, (*b*) at the magnetic equator. What is the strength of the equivalent magnet, and the total field at latitude 54°?

194. Two small magnets of magnetic moments m_1 and m_2 are placed at right angles to one another a distance d apart so that the axis-line of m_2 passes through the centre of m_1. Find the force and couple acting on each magnet. Show that these are equivalent to a single force at right angles to the straight line joining their centres and meeting that line at one-third of its length from the longitudinal magnet.

195. A simple pendulum with a period of 1 s has a bob of mass 2 g with a vertical magnetic moment of 1 A m^2. Estimate the change in the period if a similar bob, magnetized in the same sense, is placed 0.5 m below the pendulum support.

196. Find the smallest radius of curvature that can be used for the corners of a conductor charged to 100 000 V if the breakdown strength of the air is 3 V μm^{-1}.

197. A long thin wire situated at a height a above the earth carries a charge Q per unit length. What is the charge density on the ground at a point at a distance x from a line vertically below the wire?

198. A superconductor is a 'perfect diamagnetic' in the sense that $\boldsymbol{B} = 0$ everywhere inside it. Show that when a magnetized body is brought near a plane surface of a superconductor the field is distorted as if there were an image of the body, having the same polarity, equidistant from the surface on the other side.

A permanent magnet, in the shape of a cylinder 15 mm long and 5 mm in diameter of material of density 8 Mg m^{-3}, is lowered on to a plane horizontal superconductor. If its magnetic moment is 0.1 A m^2, estimate the clearance between it and the superconductor when it floats freely.

199. A positive charge Q_1 and a negative charge $-Q_2$ are separated by a distance *d*. Show that there is one spherical equipotential, and find its radius and the position of its centre, taking $Q_1 > Q_2$.

Hence find the force between an earthed conducting sphere of radius *r* and a charge Q_1 distant *a* from the centre of the sphere. ($a>r$).

200. A metal sphere of radius 0.10 m is charged to a potential of 600 V and two hemispheres made of thin metal of radius 0.11 m are brought up from a distance, without touching the first sphere, so as to form a second sphere concentric with the first. What is the potential of the first sphere relative to the earth (*a*) if the hemispheres are earthed; (*b*) if the hemispheres are insulated; (*c*) if the hemispheres are insulated, but are momentarily connected to the inner sphere by a wire carried on an insulating handle?

201. The conductors of a parallel-plate capacitor fitted with a guard-ring are made of material of coefficient of linear thermal expansion α_1. The active plate is supported from the guard-ring. The guard-ring itself is supported from the earthed plate of the capacitor by insulating material of expansion coefficient α_2. What must be the relation between α_1 and α_2 if the capacitance is to be independent of temperature?

202. The breakdown strength of air is 3 V μm^{-1}. A cylindrical capacitor consists of an outer tube whose diameter is fixed at 20 mm and an inner tube whose size can be chosen to suit the conditions of the experiment. What diameter of inner conductor should be chosen: (*a*) to make the potential difference between the inner and outer a maximum; (*b*) to make the energy per centimetre length a maximum?

203. An electroscope consists of a gold leaf *A* mounted within an earthed metal box *B*, and connected by a wire passing through a hole in *B* to a charging plate *C*. To charge it by induction another plate *D* is brought close to *C* and raised to a potential of 1 k*V*, while *C* is earthed; *C* is then isolated and *D* removed. If the capacitance between *C* and *D* during charging is four times the total of *C* to earth when *D* is absent, what is the final voltage of the gold leaf?

204. The area of the plates of a parallel-plate capacitor is 10 cm^2 and the plates are separated by sulphur of dielectric constant 4.0 and density 2 Mg m^{-3}. If four electrons per sulphur atom take part in passing a current of 10 A at a frequency of 1 MHz estimate the amplitude of oscillation of these electrons. [Atomic weight of sulphur = 32.]

205*. The plates of an air-spaced parallel-plate capacitor are a distance *a* apart and are connected externally by a wire. A charge *Q* lies between the plates. When *Q* is moved a distance *x* normal to the plates, what charge flows through the external wire?

206. The plates of a parallel-plate capacitor of large area are separated by 3 mm. The space between them is filled by a layer 2 mm thick of dielectric constant 6, and a layer 1 mm thick of dielectric constant 2. The plates are charged to a potential difference of 1 kV, the plate next to the thicker layer being positive and the other earthed. Calculate (a) the charge density on the plates, and (*b*) the potential of the interface between the dielectric layers. If the potential difference between the plates is kept at 1 kV and between the layers of dielectric is inserted a very thin sheet of insulator carrying a charge density of 3 μC m^{-2}, what will be the potential of the sheet?

207. Two capacitors of 0.01 μF each are enclosed in metal cases which approximate to spheres of 50 mm radius; one lead to each capacitor is connected to its case and the other lead emerges from the case. The two emergent leads are connected together, and a 100 V battery is connected between the cases. If the capacitors are distant from each other and from other objects, and if the negative terminal of the battery is earthed, what is the potential of the lead connecting the capacitors, and what are the charges on the capacitor plates connected to the terminals? If the connexions are reversed so that the cases are joined and the emergent leads go to the battery, what now are the charges on the capacitors, and what is the potential of the cases?

208. Two isolated spherical conductors of radii 30 and 90 mm are charged to 1.5 and 3 kV respectively. They are then connected by a fine wire. How much heat is generated in the wire?

209. A capacitor consists of two plates each 0.1 m square separated from one another by a distance of 1 mm and is charged to a potential difference of 600 V. Calculate the change in the energy stored in the condenser caused by introducing oil of dielectric constant 3 between the plates (*a*) if the condenser remains connected to the battery, (*b*) if the battery is disconnected before the oil is introduced.

210. A sulphur sphere of radius 0.1 m and dielectric constant 3.4 is uniformly charged throughout its volume to a charge density of 30 μC m^{-3}. In the course of time this charge redistributes itself as a surface charge. Calculate the loss of electrical potential energy. If the specific heat capacity of sulphur is 1.4 MJ m^{-3} K^{-1} and the sphere is thermally isolated, what is its rise of temperature?

211. Estimate the following quantities approximately:

(*a*) The force on one capacitor plate of a pair 0.1 m in diameter and 10 mm apart, charged to a potential difference of 10 kV;

(*b*) The flux density between the poles of an iron magnet, if the forces due to the magnetic field equal the atmospheric pressure;

(*c*) The force per unit length on each of a pair of parallel wires 0.1 m apart carrying a current of 1 kA;

(*d*) The self-capacitance of the earth in farads;

(*e*) The energy stored when the earth is charged so as to have a field strength of 100 V m^{-1} at all points on its surface;

(*f*) The energy stored in a 1 mF capacitor charged to 10 kV;

(*g*) The height to which a mass of 10 kg would have to be raised to equal the energy in (*f*);

(*h*) The angular velocity of a flywheel of mass 100 kg, radius of gyration 0.5 m, for its energy to be the same as in (*f*);

(*i*) The mass of water that could be raised from room temperature to boiling point by the energy in (*f*);

(*j*) The energy stored in a coil 0.1 m long, 20 mm in diameter, wound with 1000 turns and carrying 10 A;

(*k*) The velocity of an electron whose energy is 10^3 eV, and of a proton whose energy is 10^6 eV;

(*l*) The energy in eV of one molecule of a gas at 1000 K;

(*m*) The drift velocity of electrons in a copper wire carrying a current of density 10 MA m^{-2}, assuming one free electron per atom;

(*n*) The magnetic flux density at the nucleus of a hydrogen atom produced by the electron in its lowest Bohr orbit.

212. A potential difference of 3 kV is maintained between a long thin wire and an infinite conducting plane 10 mm away. What is the force per cm exerted on the wire if the capacitance per unit length between the wire and the plane is 10 pF m^{-1}?

213. A long vertical metal cylinder of inside radius 50 mm contains another long coaxial metal cylinder with outside radius 40 mm, whose ends are well inside the outer cylinder. Both cylinders are partially immersed in a liquid of dielectric constant 3.

(*a*) What change in capacitance results from lowering the inner cylinder 10 mm?

(*b*) If a potential difference of 3 kV is applied between the cylinders what is the extra vertical force on the inner cylinder?

214*. (*a*) A parallel-plate capacitor whose plate separation is a has between the plates an irregular dielectric body whose effective polarizability is K, i.e. when a voltage V is applied to the capacitor to produce a field strength E equal to V/a, the body acquires a dipole moment KE parallel to E. By considering the charge which flows in an external circuit when the dipole is set up (cf. question 205) or otherwise, show that the capacitance is increased by K/a^2 and the stored energy by $\frac{1}{2}KE^2$.

(*b*) The capacitor of (*a*) has its plates vertical and connected to a battery so that the field between the plates is E. A long dielectric rod of uniform cross-section hangs vertically with one end well inside and the other well outside. By considering the work done by the battery and the change of stored energy when the rod is moved parallel to its axis, show that the rod is attracted into the plates with a force $\frac{1}{2}KE^2$, where K is now the polarizability of the rod per unit length.

215. A long paramagnetic circular rod of diameter 5 mm is suspended from a balance so that one end is between the poles of a magnet providing a horizontal field, and the other well clear of the field. When a flux density of 1 T is applied the apparent mass of the rod increases by 1.5 g. What is the volume susceptibility of the material?

216*. (*a*) When a current of 1 A flows in a certain primary coil the flux density inside is b. A small secondary coil of n turns, each having area A, is placed in it, and is connected to a circuit which ensures that when a current i flows in the primary coil, αi flows in the secondary. Show that the effective inductance of the primary coil is increased by αnAb.

(*b*) In a coil for which $b = 10$ mT it is found that the inductance is increased by 0.8 μH because of the secondary coil. When the current in the primary is such as to produce a flux density of 0.1 mT, what is the magnetic moment of the secondary coil?

(*c*) The secondary coil is replaced by a piece of magnetizable material of volume 4 cm^3, and it is found also to increase the inductance of the primary coil by 0.8 μH. What is the volume susceptibility of the material?

217. Two spherical uniform distributions of charge, identical except in sign, are superposed with their centres not quite coincident. Show that within the region common to both spheres the resulting electric field is constant. Hence show that when a dielectric sphere is placed in a uniform field $\boldsymbol{E}$, the field within it is uniform and of strength $\boldsymbol{E}-\mathrm{n}\boldsymbol{P}/\varepsilon_0$ where n, the depolarizing factor, has a value $\frac{1}{3}$. What is the value of n for a long circular cylinder when the field is (*a*) parallel, (*b*) perpendicular to the axis?

218*. A soap bubble (the film may be taken to be a conductor) of radius 10 mm and surface tension 20 mN m^{-1} is charged by momentarily connecting it to an electrode at 6 kV. What is its increase in radius? If the uncharged bubble is placed in a uniform field of strength 150 kV m^{-1}, show that it becomes slightly ellipsoidal and determine the ratio of polar to equatorial axes.

219. A long rod of magnetic material is freely suspended in a uniform magnetic field. Show that whether it is paramagnetic or diamagnetic it tends to turn parallel to the field. What angle must the axis make with the field for the couple to be greatest?

220*. An anisotropic paramagnetic crystal of mass m is suspended from a torsion wire of torsion constant μ; the principal mass susceptibilities in the horizontal plane, in directions at right angles to one another, are χ_1 and χ_2 ($\chi_1 > \chi_2$). A horizontal magnetic field with flux density B is applied parallel to the axis of χ_1. Show that if the torsion head is now twisted steadily, the crystal also twists steadily unless $B^2 > \mu_0\mu/[m(\chi_1 - \chi_2)]$, in which case as the torsion head is twisted beyond a certain critical angle the crystal becomes unstable and suddenly jumps to a different orientation.

If $m = 10$ mg, $\mu = 10$ nN m rad^{-1}, $\chi_1 - \chi_2 = 20$ mm^3 kg^{-1}, $B = 0.5$ T, how far can the torsion head be turned before instability occurs?

221. A certain ferromagnetic material has a remanent magnetic flux density of 0.5 T and a coercivity of 50 kA m^{-1}. Assuming that part of the hysteresis loop between the remanent point and the point of zero flux density to be a quadrant of an ellipse, estimate the flux density in the 10 mm gap of a permanent magnet made from the material, the shape of which, apart from the gap, is a ring of mean radius 20 mm.

222. A thin insulating disc of radius r carries a total charge Q uniformly distributed on its surfaces. Find the magnetic flux density produced at a point on the axis of the disc, distant x from its centre, when the disc is rotated with angular velocity ω?

223. A non-magnetic toroid, consisting of a circular bar bent into a circular ring, is uniformly wound with N turns of wire carrying a current I. What is the magnetic flux density at any point inside or outside the toroid? If the cross-section of the bar from which the toroid is constructed is uniform but not circular, how is the answer affected?

224. Two similar parallel coaxial circular coils of radius a and separation b carry a current I. Find the value of b that makes the variation of the magnetic field along the axis of the coils near the point midway between the coils as small as possible. Show that with the correct separation the flux density $B(x)$ on the axis of the coils at a distance x from the midpoint is such that $[B(x) - B(o)]/B(o)$ is proportional to x^4 when $x \ll a$.

225. A uniform solenoid of diameter 0.1 m is to be designed so that the field on the axis does not decrease by more than 1 % in moving from the centre to a point 10 mm away along the axis. How long must the solenoid be?

226*. In the previous question if the field along the axis varies as $a + bx^2 + cx^4 + \ldots$ near the centre, x being the distance from the centre, show that the term bx^2 can be eliminated by passing no current through a short length of the windings near the centre, and determine this length.

227*. A spherical non-magnetic shell is wound with wire in such a way that there are 2 turns in every millimetre measured in a direction normal to the plane of the windings. A current of 1 A is passed through the wire. Show that the resulting field is the same at all points within the shell, and determine the flux density.

228. A transmission line consists of two long conducting strips of width 0.1 m, facing each other and 5 mm apart, and carrying the same current in opposite directions. What are the inductance L and C of the system per unit length? Determine also the product LC and show that it is independent of the dimensions of the line.

229. Two coplanar concentric circular rings have radii 0.5 m and 10 mm. Show by direct calculation that the flux linked with one ring when unit current is passed round the other is the same whichever ring carries the current. What is the mutual inductance of the two rings?

230. Two parallel wires 1 m long and of mass 100 g are suspended 0.1 m apart from supports at the same level in such a way that the sag at the midpoint of each wire is 50 mm. In what directions and how far will the midpoints of the wires move if currents of 100 A are passed through both wires in the same direction?

231. When a homogeneous beam of electrons is passed through an evacuated region where there is simultaneously present an electric field of 30 V mm^{-1} and a magnetic flux density of 3 mT it is found that the electrons are not deflected. When the magnetic field alone is present the electrons move in a circle of 19 mm radius. Calculate the velocity of the electrons and the ratio of their charge to mass. (The electric and magnetic fields are at right angles to one another and to the path of the electrons.)

232. An oscillograph consists of a phosphor bronze wire 0.14 m long of which the middle 40 mm lies in, and at right angles to, a magnetic field of flux density 0.6 T. If the wire is kept taut under a tension of 0.1 N calculate approximately the current which will give a deflexion of 1 mm.

If the density of the wire is 9 Mg m^{-3} and its radius 30 μm, and if the wire is suitably damped, up to about what frequency would the oscillograph be reasonably sensitive to an alternating current?

233. Two long solenoids of radius 20 mm, wound with 2 turns/mm, are placed end to end, nearly touching. When 1 A is passed in the same sense through each solenoid, what is the force between them? Is it attractive or repulsive?

234. A large solenoid, 0.4 m in diameter and many metres long, is to be constructed from superconducting wire which can pass a huge current without resistance. If the winding consists of 400 turns/m of wire 2 mm in diameter, whose ultimate tensile strength is 2×10^8 N m^{-2}, and if no additional strengthening is provided, what is the maximum current that can be passed and what flux density is attainable without bursting the coil?

235. A soft iron anchor ring of mean radius 0.1 m, of uniform cross-sectional area 500 mm^2, and with a gap 1 mm long is wound with 200 turns of wire carrying 5 A. If the B–H relation for the iron is given by the table below, what is the field in the gap? If the current is reduced linearly to zero in 1 ms what voltage is developed across the ends of the winding?

H/A m^{-1}	50	100	150	200	300	500	800	1500
B/T	0.1	0.3	0.55	0.73	0.95	1.2	1.4	1.5

236. A network is formed from uniform wire in the shape of a rectangle of sides $2a$, $3a$, with parallel wires arranged so as to divide the internal space into six squares of side a, the contact at the point of intersection being perfect. Show that if a current enters the framework at one corner and leaves it by the opposite corner, the resistance is the same as that of a length $121a/69$ of the wire.

237. Two cells of electromotive force e_1 and e_2, and resistance r_1 and r_2, are connected in parallel to the ends of a wire of resistance R. Show that the current in the wire is

$$(e_1r_2+e_2r_1)/(r_1R+r_2R+r_1r_2)$$

and find the ratio of the work done by each cell, excluding what is dissipated as heat in the cells.

238. A wire mesh is composed of two sets of parallel wires welded together so that each cell of the mesh is a non-rectangular rhombus. A long strip is cut from the mesh parallel to one set of wires, and a current is passed along the strip. What is the form of the equipotential lines at some distance from the ends of the strip?

239. A Wheatsone bridge is used to find the melting point of a substance by observations on the temperature variation of the resistance of a coil embedded in the substance. If the resistance of the galvanometer and of each element of the bridge at the melting-point is 100 Ω, the e.m.f. of the battery is 0.1 V, and the smallest perceptible change of galvanometer current is 10^{-8} A, find the accuracy with which the melting point can be determined, taking the temperature coefficient of resistance of the thermometer wire to be 4 kK^{-1}.

240. The magnitude I of the current in a non-ohmic resistance unit is given by $I = 0.004V^4$, where i is in amperes and V is the applied potential difference in volts. The unit is connected in one arm of a Wheatstone bridge and each of the other arms has a resistance of 16 Ω. The current supplied from the battery is adjusted until the bridge is balanced. Find the current necessary for this.

241. An apparatus is earthed by means of a hemispherical metal electrode 0.2 m in diameter which is sunk in the earth in such a way that the circular base is flush with the earth's surface. If the resistivity of the earth is 100 Ω m, what is the earthing resistance?

242. An air-spaced 100pF capacitor is immersed in a fluid whose resistivity is 10 Ω m. What is the resistance between its plates?

243. A nichrome disc of 0.1 m radius and thickness 2 mm is mounted on an axle of 10 mm radius and rotated at 3000 r.p.m. in a uniform magnetic field of 0.2 T. Brushes, connected in parallel, make contact with many points on the rim of the disc, and a similar set of brushes touches the axle. Calculate the maximum power in watts available in an external circuit connected to the two sets of brushes.

[Resistivity of nichrome = 1.1 $\mu\Omega$ m.]

244. In order to calibrate an electromagnet, a search coil of 20 turns is wound on a cylindrical former so that the mean diameter of the windings is 10 mm. The coil is connected in series with the secondary of a 1 mH mutual inductor and a ballistic galvanometer. It is found that the same ballistic throw is obtained by removing the search coil from the field as by reversing a current of 0.823 A in the primary of the mutual inductance. What is the strength of the field?

245. Show that the magnetic energy of a coil of finite size carrying a given current is finite, but increases, as the radius r of the wire is decreased, towards infinity as $\ln(1/r)$.

246. If the flux density at the centre of a long uniform solenoid is B when unit current is passed, show that the inductance is approximately VB^2/μ_0, where V is the volume of the solenoid. Estimate the error for a single layer solenoid whose length is 5 times the diameter.

247*. A circular ring, made of wire whose cross-sectional area is 5 mm^2, is mounted so that it can be spun about a vertical diameter. A small damped compass needle is placed at the centre of the ring. With the ring at rest the needle points along the direction of the earth's field, and it is found that when the ring is spun at 6000 r.p.m. the needle turns through 1 degree to a new position of rest. What is the resistivity of the material of the wire?

248. Stressed permalloy has only two stable magnetic states, and thus has a rectangular hysteresis loop lying between the limits $\pm$ 200 A m^{-1} for the magnetising field, and $\pm$ 0.1 T for the flux density. Calculate the rates of temperature rise due to hysteresis of a specimen when placed in a 50 Hz alternating magnetic field $H_0 \sin \omega t$ for the cases in which H_0 is 100, 300 and 500 A m^{-1}.

[Specific heat capacity of permalloy = 0.40 J g^{-1} K^{-1}, density = 8 Mg m^{-3}.]

249. A moving coil galvanometer consists of a coil of 70 turns each 50 mm^2 in area suspended in a radial field of 0.3 T by a fibre whose torsion constant is 6 nN m rad^{-1}. The moment of inertia of the suspended system is 0.8 μg m^2 and its resistance 25 Ω. To observe its movement a mirror 4 mm wide is attached to the coil, the associated lamp and scale being 1 m away. Air damping of the suspension may be ignored. Calculate the following properties:

(*a*) The current senstivity in mm deflexion per μA.

(*b*) The quantity senstitivity in mm deflexion per μC on open circuit.

(*c*) The time of swing on open circuit.

(*d*) The minimum width of the image of a fine cross-line produced by the optical system.

(*e*) The external resistance for critical damping.

(*f*) The resistance of a single loop of wire 50 mm^2 in area, mounted on the coil former, which will produce critical damping on open circuit.

(*g*)* The quantity sensitivity when critically damped.

(*h*) The ratio of successive swings in the same direction with 2 kΩ in the external circuit.

(*i*) If the suspended system has Brownian motion amounting to $\frac{1}{2}kT$ per degree of freedom, what is the r.m.s. amplitude of the motion of the light-spot? If this marks the limit of sensitivity, what is the smallest detectable current?

250. A moving coil galvanometer is to be modified for use as a detector of small thermoelectric voltages in a circuit of resistance R (excluding the galvanometer resistance). The modification consists of rewinding the coil with wire of a different diameter, the overall dimensions of the coil and its mass being kept the same as before. Show that the greatest sensitivity is achieved when the wire is chosen so that the galvanometer resistance is R. Show also that this is the condition for dissipating most power in the galvanometer.

251. The terminals of a source of constant potential are connected to the ends of a coil of large self-inductance and small resistance. Show that at first the current is independent of the resistance, but that ultimately it is dependent only on the resistance.

252. Two inductors L_1 and L_2 carry steady currents I_1 amd I_2 respectively, and have mutual inductance M. Obtain an expression for the stored energy in the circuit, and show that for this to be always positive L_1 and L_2 must be positive and $M^2 \leqslant L_1L_2$. If L_1 and L_2 are connected in series, show that the combination can only have zero inductance if $L_1 = L_2 = M$.

253*. A bank of capacitors of total capacitance 2 mF is charged to 2 kV; and then connected across the terminals of a small coil which is cooled so as to have a very low resistance. Estimate the maximum magnetic flux density generated in the coil, if the volume over which the field strength is comparable with the value at the centre is 20 cm³. If the inductance of the coil is 5 mH, how long after the connexion does the field attain its maximum value, and what is then the current in the coil?

254. A coil consists of 500 turns of radius 10 mm, uniformly wound along a length of 0.25 m. A capacitor *C* is connected across the coil. If the natural frequency of the circuit is 25 kHz what is the approximate value of *C*? (Neglect end effects in the coil and assume its resistance is zero.)

255*. It is desired to produce a magnetic field of peak amplitude 20 mT, oscillating at about 1 kHz, and for this purpose a solenoid is constructed 0.5 m long, 80 mm internal diameter, uniformly wound with copper wire (resistivity 18 nΩ m) to a thickness of 10 mm. Of the volume of the windings, 15% is occupied by insulation, 85% by copper, and the total resistance is 30 Ω. A capacitor is connected across the solenoid to give resonance at the required frequency and the two are driven in parallel by an oscillator. What approximately is the required capacitance, and what maximum voltage must it withstand? What is the r.m.s. current delivered by the oscillator? If the oscillator current is constant, over what range of frequency will the amplitude of the field be within 10% of its maximum?

256. A coil which is to be used as a standard resistor at 1 kHz has a d.c. resistance of 1000.000 Ω and a self-inductance of 10μH. What is its phase angle at 1 kHz? Show that it is possible to reduce the phase angle to zero at this frequency by putting a capacitor *C* in parallel with the coil, without introducing any significant change in the apparent resistance of the coil. What is the value of *C*, and over what range of frequencies is the phase angle now numerically less than it was before *C* was included in the circuit?

257. A box provided with two terminals is known to contain an inductor of very small resistance, a capacitor and a resistor. When 1 V d.c. is connected to the terminals, 1 mA flows. When 1 V a.c. at 50 Hz is connected, 10 mA flows. If the frequency is increased and the applied voltage maintained constant, the current rises to a very high maximum at 1 kHz. As the frequency is increased further the current drops steadily, tending ultimately to 1 mA. How are the components connected inside the box and what values do they have?

258. Two circuits in parallel consist of a resistor of 100 Ω in series with an inductor of 0.1 H and a resistor of 200 Ω in series with a capacitor. What must be the capacitance for the currents in the two circuits to be in phase quadrature? At what frequency are the currents numerically equal?

259. An ideal transformer consists of two inductors L_1 and L_2, whose mutual inductance M is $(L_1L_2)^{\frac{1}{2}}$ and whose resistance is zero. If a

resistor R is connected across the terminals of L_2, what is the impedance between the terminals of L_1 at a frequency ω such that ωL_1 and ωL_2 are both much greater than R?

260. Two identical resonant circuits each consist of a coil having an inductance of 500 μH and a resistance of 25 Ω, tuned by a total capacitance of 500 pF. An e.m.f. of 10 V at the resonant frequency of the individual circuits is injected in series in the primary circuit.

Find the mutual inductance between the coils which would produce critical coupling (maximum secondary current) and find the value of the secondary current for this coupling.

If the mutual inductance is 25 μH estimate the frequency (or frequencies) at which the secondary current is a maximum.

261. A cycle-hub dynamo consists of a four-pole permanent magnet which rotates with the front wheel, and a fixed armature. Over the four poles of the armature are wound coils which are connected in series and in series with the bulb of the cycle lamp. Show that if the self-inductance of the coil system has a sufficiently high value, the current taken by the bulb can be made nearly independent of speed.

Such a dynamo fully lights a 1.5 W 6V bulb at a cycle speed of 10 m.p.h. and gives 4 V on open circuit at 4 m.p.h. The diameter of the wheel is 26 in. Estimate the self-inductance of the dynamo and the current through the bulb at 5 and 15 m.p.h. (Ignore the resistance of the dynamo and assume that the resistance of the bulb is constant.)

262. Show that if a small alternating e.m.f. is applied to the galvanometer of question 249, the current that flows is the same as would flow through a certain capacitor in series with the clamped galvanometer coil. Determine the value of the capacitor on the assumption that the alternating frequency is much higher than the natural vibration frequency of the galvanometer.

263. An alternating current meter consists of a full-wave rectifier and a direct-current moving-coil meter. The instrument is calibrated so as to read r.m.s. current correctly when used to measure a sinusoidal current. By how much will the indicated r.m.s. current be in error when it is used to measure (*a*) a 'square-wave' current which is alternately $+I$ and $-I$ for equal periods of time, (*b*) a current with a second harmonic represented by

$$I = I_0 \cos\omega t + 0.2 I_0 \cos 2\omega t\,?$$

264. When a voltage V_g is applied between the grid and cathode of a triode, the anode current $I = a + bV_g + cV_g^2$. If V_g consists of a steady bias and an oscillatory signal, $V_g = A + B \cos \omega t$, show that the mean value of the anode current contains a term proportional to B^2 (i.e. there is rectification of the oscillatory signal), and that there is a second harmonic component in the anode voltage. If I_0 is the value of I when $B = 0$, and when the signal is applied I oscillates between I_{max} and I_{min}, express the ratio of amplitudes of the second harmonic and fundamental in terms of the quantity X, defined as $(I_{max}-I_0)/(I_0-I_{min})$.

265. A discharge tube, which becomes conducting when the potential difference across it reaches 100 V and extinguishes its discharge if the potential difference falls below 50 V, is connected across a 1 μF capacitor. Describe quantitatively what happens if a 500 V battery and a 1 MΩ resistor are connected in series across the capacitor.

ANSWERS

1. $\surd 5$. **2.** $\frac{5}{3}, \frac{5}{4}$ nautical miles.

3. If L is the line joining the midpoints of AC and BD, a force $\surd 2$ along L and a torque $\frac{1}{2}$ with L as an axis.

4. All equal. **5.** $d<r$. **7.** $\mu l(\mu+\cot\alpha)/(1+\mu^2)$.

8. $2\pi rF(e^{2\pi n\mu}-1)$. **9.** 38, 132 ft.

10. (*a*) 3 tons weight. (*b*) 30 kW. (*c*) 2.6×10^{12} J; 12 h. (*d*) 60 ft.

11. (*a*) 5.7×10^{-13} m; they exchange velocities.
(*d*) 18.2×10^{6} m s^{-1}.

12. At V ft s^{-1}, thrust $= 1.624\times 10^5 - 100V$ ft lb s^{-2},
power $= 7280 - 2.8\times 10^{-3}V^2$ h.p.; max. thrust $= 1.624\times 10^5$ ft lb s^{-2},
max. power $= 7280$ h.p., max. speed $= 1624$ ft s^{-1}.

13. 123 km. **14.** $\dfrac{Ft}{m_0}\log_e\left(\dfrac{m+m_0}{m}\right)-gt$.

16. $2m/M$; 7.53 km s^{-1}, 94 min; 1.48 km; 10^{-11} kg m^{-3}.

17. 3.6. **18.** (*c*) 53, 1.1 pm; 13.4, 33 500 eV.

21. 500 mm to south; 500 mm to south and 75 mm to east.

24. $gr^2/(r^2+k^2)$; $mgk^2/(r^2+k^2)$.

25. Uniform rotation with angular velocity ω about centre, which has upward velocity $\frac{1}{6}\omega l - gt$.

26. $\dfrac{b(4b+5h)}{6(b+h)}$ from the hinge. **27.** $4\omega/5$; 1/5.

28. 196 km h^{-1}. **29.** 0.245 rad s^{-1}.

32. The angle between $\boldsymbol{B}$ and $\boldsymbol{M}$ oscillates with a frequency which is the difference between γB and the frequency of b.

34. 223 Hz. **35.** $r/\surd 2$ from the centre.

37. (i) $T = \surd(3/2)T_0$, amplitude and energy unchanged.
(ii) $T = \surd(3/2)T_0$, $A = \surd(2/3)A_0$, $E = 2/3E_0$.

38. $2\theta(T/I)^{\frac{1}{2}}$. **43.** 1.13.

47. The oscillation only affects a few masses near the point of excitation.

49. 0.032. **50.** $l/9$; $W/8$.

52. (*a*) 1/9; min. (*b*) 1.27%; min. (*c*) 0.009%; max.
(*d*) 17.4%; max.

53. $(F\rho)^{\frac{1}{2}}$. **54.** 38%; 14.4 mm. **56.** 299 799.8 km s^{-1}.

57. (*a*) $\frac{1}{2}$. (*b*) 3/2; 1.08 $(\gamma g/\rho)^{\frac{1}{2}}$. **59.** 2500 K.

60. $\frac{1}{2}$. **61.** 7.5 km s^{-1}; 400 km.

63. A descending tone of frequency $\nu_0 t/(t^2-t^2)^{\frac{1}{2}}$, where $\nu_0 = v/2a$ and $t_0 = 2D/v$, v being the velocity of sound in air.

64. $\frac{1}{2}v/V\frac{1}{3}$.

66. $n = 1.5-0.4r^2$; $f = 5$ m.

67. nd

68. 2.97 mm.

69. 22; 14.

70. 1600 K.

71. 121 μm, 180.

72. (*a*) 500 μm. (*b*) 15–20. (*c*) 0.152 m. (*d*) 20 μm.

73. 0.546 μm.

74. 96.594 ± 0.006.

75. $(R_1+R_2)(pn\lambda/d)^{\frac{1}{2}}$, p integral.

77. (*a*) Straight fringes 0.15 mm apart.
(*b*) Images of the pinholes 10 mm apart, the images having diameter 0.44 mm to first zero of diffraction pattern.

78. (*a*) 2 mm. (*b*) 1 mm. (*c*) 1 mm.

79. (*a*) Along a line normal to the slit, at the centre and at distances $(n+\frac{1}{2})$ mm, n being an integer > 0.
(*b*) The same as in (*a*), but the centre is much brighter.
(*c*) Everywhere on a rectangular grid of lines 0.1×1 mm, the central lines of the grid being absent.
(*d*) Everywhere on lines normal to the line of the holes, $n/3$ mm from the centre (n not a multiple of 3).
(*e*) At all vertices of a hexagonal net (a honeycomb) with sides $\frac{1}{3}$ mm long parallel to the sides of the triangle.

80. 0.595 μm; $-32°$.

81. 5° 44′, 11° 32′, 17° 28′, 23° 35′, 30°; 0.686, 0.171, 0, 0.043, 0.027 (neglecting obliquity factor); 4″.

82. Intensity 0.64 at 0°, 4°, 8°, intensity 0.04 at 0.8°, 1.6°, 2.4°, 3.2°, 4.8°, 5.6°, 6.4°, 7.2°.

83. $1.000\,294\,2 \pm 0.000\,000\,2$.

84. 0.1, 0.01 seconds of arc.

85. 50 mm; 100 m.

86. 4000.

87. 17.4 mm; 982; 1320 lines/mm.

88. 2 pm, 0.64, 1° 42′.

89. 2.0, 0.67, 0.40, 0.29 m, etc.

90. 0.25 m, 20.9 mm (convex).

91. 1.1 mm; 0.4 mm.

92. 0.11m, 2.5 mm; 0.155 m, 2.5 mm.

93. 1.80, 52°.

94. 11.95 mm.

95. $2.5 \times 10-^5$; $\frac{5}{3}$.

96. 2.5 mm.

97. 200 Hz.

98. 2.4×10^{11} units; 10.9 units; 4300.

99. 3×10^{10} cm; 7.4×10^{-29} cm.

102. 27 mN m^{-1}.

103. The same.

104. 1.4×10^6 atm.

105. 2.4 Mg m^{-3}.

106. 0.97 mg.

108. (*a*) $W+v\rho_2 g$. (*b*) ? (*c*) $W+v\rho_1 g$.

110. 1.12×10^{-2} m³ s^{-1}.

111. 21 r.p.m.

112. 612 cm³ s^{-1}, 18.4 mm s^{-1}.

113. 38 s.

114. 9.5×10^7 s.

115. 1.36r.

116. 1.61×10^{-18} C.

117. There is an error in the third result of set (*b*); 5.94 m s^{-1}.

120. 4.9 μm.

121. 2.72 kg.

122. 37° off vertical.

123. 1.5×10^{11} N m^{-2}. **126.** 0.354 m.

127. (*d*) 5×10^{-5} s; 0.1 m.

128. $r_0/\alpha - 2\gamma/(\rho g r_0)$, $2\gamma/(\rho g r_0)$; the lower; the liquid rises to the top.

129. The smaller blows up the larger until both have the same radius; they become identical.

130. 9.2 °C; 1.36; 3.7 nm. **131.** 16.95 mN m^{-1}.

132. 1.54 K. **133.** 8 mm. **134.** 2.2×10^7 s.

135. (*a*) 235 J. (*b*) 188 J. (*c*) 275 J.

136. 15 h. **137.** About 5 mm.

140. 24 g s^{-1}; reduced by factor 83.

141. 2.2×10^{-3} m^3 s^{-1}; 365 K. **142.** 1.68; ± 0.397 m.

143. $\gamma = 19/13$; 1.38 K, 20. mmHg.

144. (*a*) $2^{-\frac{2}{3}}T$. (*b*) $1.4T$. (*c*) T; $\frac{5}{3}T$. (*d*) T. (*e*) $0.758T$; $1.47T$.

145. 5.0 mm.

146. (*a*) 53 kK m^{-1}; (*b*) 9.8 mK m^{-1}; convection.

147. 34 atm.

149. (*a*) Up to 369 °C the pressure equals the vapour pressure; above 369 °C it is roughly proportional to absolute temperature; there is a sharp change of gradient at 369 °C.

(*b*) The thermal capacity rises sharply to a high but finite value at 369 °C, where it drops discontinuously to a low value, thereafter staying nearly constant.

150. 14.5 atm. **151.** 326.6.

152. 771 mmHg; 21 mmHg. **153.** 15%.

154. 140 kK.

157. (*a*) 0.03 s. (*b*) 1.85×10^{-4} mmHg. (*d*) 1.82×10^{-3} mmHg.

158. (*a*) $(8RT/\pi M)^{\frac{1}{2}}$; $(3RT/M)^{\frac{1}{2}}$; $(2RT/M)^{\frac{1}{2}}$; $\frac{3}{2}kT$.
(*c*) 88 mW m^{-1}; 3.2×10^{-2} mmHg.

159. (*a*) 1.34. (*b*) 0.52, 3.0. **161.** 5×10^{-2} mmHg.

162. (*a*) 5/3. (*b*) 9/7. (*c*) 15/13. (*d*) 7/6. (*e*) 34/33.

163. (*a*) 1.04×10^{-21} J; 50 K. (*b*) 6.1×10^{-20} J; 1000 K.

165. 300 pm; 10^{-21} J. **166.** 7.7. **167.** 620 K.

168. (*a*) 2.5 atm, 2.53×10^{-3} m^3, and 1 atm, 6.32×10^{-3} m^3.
(*b*) 928 J, 586 J; 36.9%. (*c*) Efficiency now 31.2%.

169. 3/31.

170. (*a*) 5.0×10^{10} J. (*b*) 2.6×10^{10} J. (*c*) 60.5×10^{10} J.
(*d*) 1.42×10^9 J. (*e*) 1.19×10^9 J.

171. (*a*) 6.4 J K^{-1}. (*b*) 1.38 J K^{-1}. (*c*) 3.6 J K^{-1}.
(*d*) 1.83×10^{-5} J K^{-1}. (*e*) 1.83×10^{-5} J K^{-1}.
(*f*) 0. (*g*) 0. (*j*) 5.76 J K^{-1}.

172. 400 K.

176. Ratio $= \dfrac{\gamma}{\gamma-1}\left\{\left(\dfrac{p_1}{p_0}\right)^{(\gamma-1)/\gamma}-1\right\}\Big/\ln\left(\dfrac{p_1}{p_0}\right)$.

177. 95 atm. **178.** 31 atm; 19%.

179. (*a*) 27/4. (*b*) 8.

180. 25.5 K; 49 K. **181.** 0.13 K.

183. 121.4 mJ m^{-2}, 0.167 mJ m^{-2} K^{-1}; 36.75 °C; 13.7 mJ K^{-1}.

184. 327.75 °C.

185. dL/dT computed from vapour pressure data $= -2.1$ J g^{-1} K^{-1}; from c_p data $dL/dT = -2.3$ J g^{-1} K^{-1}. At 20 °C $L = 2460$ J g^{-1}.

186. 4.8 eV; 257 nm. **187.** 35.3 atm. **188.** 18.6, 23.9 J g^{-1}.

190. 12.3 mV, 29.1 μV K^{-1}; 526 °C. **192.** 59 mm.

193. (*a*) 62 μT, vertical.
(*b*) 31 μT, horizontal; 7.9×10^{22} A m^2, 53 μT.

194. Force between magnets $= 3\mu_0 m_1 m_2/(4\pi d^4)$
Torque on $m_1 = \mu_0 m_1 m_2/(2\pi d^3)$; on $m_2 = \mu_0 m_1 m_2/(4\pi d^3)$.

195. -11.5 ms. **196.** 33 mm.

197. $Qa/\pi(a^2+x^2)$. **198.** 5 mm.

199. $Q_1Q_2d/(Q_1^2-Q_2^2)$, $Q_2^2d/(Q_1^2-Q_2^2)$ from negative charge; $Q_1^2r/[4\pi\varepsilon_0(a^2-r^2)]$.

200. (*a*) 600/11 V. (*b*) 600 V. (*c*) 6000/11 V.

201. $\alpha_2 = 2\alpha_1$. **202.** (*a*) 7.36 mm. (*b*) 12.1 mm.

203. 4 kV. **204.** 5×10^{-14} m. **205.** Qx/a.

206. (*a*) 10.6 μC m^{-2}; (*b*) 600 V; (*c*) 668 V.

207. 50 V, $\pm0.5\mu$C; 0.500 28 and -0.499 72 μC; 49.972 V.

208. 2.6 μJ. **209.** (*a*) 31.8 μJ. (*b*) -10.6 μJ.

210. 41.8 μJ; 7.1 nK.

211. (*a*) 35 mN. (*b*) 0.5 T. (*c*) 2 N m^{-1}.
(*d*) 7×10^{-4} F. (*e*) 1.4×10^{14} J. (*f*) 5×10^{4} J.
(*g*) 500 m. (*h*) 63 rad s^{-1}. (*i*) 150 g.
(*j*) 0.2 J. (*k*) 19 and 14 Mm s^{-1}. (*l*) 0.13 eV.
(*m*) 0.7 mm s^{-1}. (*n*) 13 T.

212. 0.81 mN m^{-1}. **213.** 5 pF; 2.25 mN.

215. 1.9×10^{-3}.

216. (*b*) 800 nA m^2. (*c*) 2.5×10^{-3}.

217. (*a*) 1. (*b*) $\frac{1}{2}$. **218.** 53 nm; 1.112.

219. 45°. **220.** 163°. **221.** 0.42 T.

222. $\dfrac{\mu_0\omega Q}{2\pi r^2}\left[\dfrac{r^2+2x^2}{(r^2+x^2)^{\frac{1}{2}}}-2x\right]$ along the axis.

223. At distance r from the axis $B = \mu_0 NI/(8\pi^2 r)$ inside the toroid, 0 outside; the cross-section is irrelevant. Strictly one should also consider the flux due to any current component circulating round the toroid.

224. $b = a$. **225.** 0.12 m. **226.** 13.4 mm. **227.** 1.68 mT.

228. 6.28×10^{-8} H m^{-1}, 1.77×10^{-10} F m^{-1}; 1.11×10^{-17} s^2 m^{-2}.

229. 3.95×10^{-10} H. **230.** 1 mm inwards.

231. 10 Mm s^{-1}; 1.75×10^{11} C kg^{-1}.

232. 0.12 A; 500 Hz. **233.** 3.15 mN, attractive.

234. 3540 A; 1.78 T. **235.** 1.1 T; 110 V.

237. $\dfrac{(R+r_2)e_1 - Re_2}{(R+r_1)e_2 - Re_1}$.

238. Parallel to the other set of wires. **239.** ± 20 mK.

240. 0.313 A. **241.** 159 Ω. **242.** 0.885 Ω. **243.** 119 W.

244. 830 kA m^{-1}. **246.** Overestimated by about 7%.

247. 28 nΩ m.

248. 0; 1.25 mK s^{-1}; 1.25 mK s^{-1}.

249. (*a*) 350 mm/μA. (*b*) 960 mm/μC. (*c*) 2.30 s.
(*d*) 0.2 mm. (*e*) 226 Ω. (*f*) 0.051 Ω. (*g*) 352 mm/μC.
(*h*) 0.455. (*i*) 1.6 μm; 4.5 pA.

253. 22 T; 5 ms; 1260 A. **254.** 0.1 μF.

255. 0.07 μF; 4 kV; 14 mA; ± 6 Hz.

256. 0.0036°; 10pF; up to 15.8 MHz.

257. 1000 Ω in parallel with (802 μH and 31.6 μF in series).

258. 5 μF; 318 Hz. **259.** L_1R/L_2.

260. 12.5 μH; 0.2 A; 311 and 325 kHz.

261. 1.18 H; 0.174 A; 0.279 A. **262.** 730 μF.

263. 11% high; 0.08% low. **264.** $\dfrac{1}{2}\,\dfrac{X-1}{X+1}$.

265. The voltage across the capacitor performs a saw-tooth oscillation, rising nearly linearly from 50 to 100 V and dropping abruptly to 50 V, with a frequency of 8.49 Hz.

INDEX

(*The numbers refer to the problems*)